The Robot Revolution

The picture depicts the author (far left) at the small manufacturing firm Perrey Dugmore Ltd in Birmingham, where as a young man he worked as an engineer in the mid 1960s

The Robot Revolution

Understanding the Social and Economic Impact

John Hudson

Department of Economics, University of Bath, UK

Cheltenham, UK • Northampton, MA, USA

Published by
Edward Elgar Publishing Limited
The Lypiatts
15 Lansdown Road
Cheltenham
Glos GL50 2JA
UK

Edward Elgar Publishing, Inc.
William Pratt House
9 Dewey Court
Northampton
Massachusetts 01060
USA

A catalogue record for this book
is available from the British Library

Library of Congress Control Number: 2018967524

This book is available electronically in the **Elgar**online
Economics subject collection
DOI 10.4337/9781788974486

ISBN 978 1 78897 447 9 (cased)
ISBN 978 1 78897 448 6 (eBook)

Typeset by Servis Filmsetting Ltd, Stockport, Cheshire
Printed on FSC approved paper
Printed and bound in Great Britain by Marston Book Services Ltd, Oxfordshire

Dedication

For Effie and Rory

Contents

Figures

Tables

Foreword

This book has developed gradually since 2015 as a result of John Hudson's research and teaching interests in robotics. Hudson approaches the rise of robots from a macroeconomic perspective, paying particular attention to the implications for employment, unemployment and wages. He also examines social attitudes to robots, both within specific labour market contexts and outside the workplace. The central theme is that robots, both within the home and in society, offer great opportunities and benefits but also substantial risks. So how does society trade these off? Hudson answers this question by presenting the first research using regression analysis that examines attitudes to robots in areas that have implications for both employment and quality of life: care for the elderly, education, surgery, driverless vehicles and dangerous situations. He finds that, on balance, people tend to be slightly in favour of robots, but this does depend upon their use. It would appear that people are balancing risks and gains, and that such risks and gains differ between uses. In general, the old, the uneducated, women and those who live outside large towns and cities, particularly those in rural areas, tend to be the most vociferous opponents, as are those in manual jobs. It is the young, male city dweller who emerges as the most accommodating.

Somewhat worryingly, Hudson shows that attitudes to robots in general are becoming slightly less supportive over time. Indeed, even amongst the well educated there is a deep well of scepticism regarding their potential use, which is unlikely to be the reflection of ignorance. And opinions tend to be polarised – many people are either vehemently opposed to or fervently in favour of robots – which is not helpful to policy-makers seeking to find a middle ground.

Hudson's approach throughout the book is rooted in his long-running interest in the macroeconomic effects of labour markets that surfaced in the 1980s when UK unemployment was at a 50-year high. This interest in contemporary employment issues facilitated his 1988 work, *Unemployment After Keynes: Towards A New General Theory*. In this, Hudson explored one explanation offered at the time – that new technology and automation meant fewer jobs for workers – which has particular resonance with his latest work.

The book speaks to the current debate over how society should respond to the rise of robots in industry and increasingly parts of the service sector. Now robots are not only used routinely in the manufacture of vehicles, but also to carry out complex operations in hospitals. The book also encompasses the parallel rise of artificial intelligence, which enables robots to adapt optimally to a given situation, and explores the ethical question as to whether the rise of robots is for the good of society. There are two contrasting viewpoints here. The optimists believe that robots will complement existing working environments and practices, increasing levels of productivity, wealth and welfare to the benefit of humankind across the world. The pessimists, towards whom Hudson is more inclined, feel instead that increased use of robots would lead to inequality and high unemployment, especially among low-skilled workers. This would lead to wealth being increasingly concentrated among those who own and operate the robots whilst leaving the low-skilled unemployed increasingly worse off. Some argue that this is not a new debate and that a fear of technology can be observed through history – for example, the nineteenth century Luddites who destroyed the textile machines that they correctly feared would replace their role in the industry. But is this time different? The question now is whether robots will replace humans in jobs across the spectrum, from street cleaners to university lecturers.

Although the book does not cover the specifics of the underlying technology, it discusses the intuition and implications of the technologies used. It also covers the history of robots and anticipates likely future changes and developments. The book concludes with a discussion of the potential problems that could arise from the advances in robotics and offers some solutions.

Although robots and their interaction with the economy is a relatively new area of economics, Hudson's interest goes back many years. In 2005 he was one of the organisers of a forum in Barcelona for the Institute of Electrical and Electronics Engineers (IEEE) on Robotics and Automation. He also encountered some of the problems associated with robotics in his work as the Vice President of the European Academy for Standardization, between 2009 and 2012.

John Hudson died in July 2018 having just completed this book. Over a long career as an economist, largely at the University of Bath in the UK, he contributed to many different areas of economics, in terms of his research and teaching. He was catholic in his interests and his work spanned beyond economics and finance to such disciplines as political science and technology standardisation. He worked on such disparate areas as bankruptcy, development, citation indexes, institutional trust and, latterly, robots. He published over 80 academic papers and a number of books and policy documents. He began lecturing at the University of Bath in 1978 and achieved a Chair in Economics in 2002. He had extensive links with universities throughout the world, especially in Slovakia.

Bruce Morley
John G. Sessions

Preface

Robots, which in this preface we will regard as including artificial intelligence, offer tremendous hope for humankind. They open the way for a world where the elderly are cared for, where the infirm can walk, where everyone can drive into the city and where people living in remote areas can enjoy many of the benefits of living in a town. A world where robot sentries guard your home and do the garden and you swallow a robot surgeon who operates on your stomach. They also open the door for increased prosperity for all, together with enhanced leisure time. And yet people are concerned that in the end robots will not lead to this vision of nirvana, but create inequality and unemployment; that they will help to destroy democracy and open the way to a surveillance state, increased criminality and lead to robot wars. Some even fear that they spell the end of human existence as we know it, either because the machines take over or because exoskeletons and brain implants change us to the point we are not recognisably human. Regardless of the extent to which either of these scenarios is true, robots will dramatically change our lives and will in the process inevitably change us.

Robots are clearly different to any other technological revolution we have had. Previous innovations provided new products or new methods of production. In doing the latter they made humans more efficient. To an extent, robots do all that, but they also replace humans in both the workplace and the home and they do this across virtually the whole range of human activity. There is simply no precedent for this. No innovation in history has been so anticipated in fiction and it is here that we see much of the worst of what might happen with the robot story. Many academics regard these fears as exaggerated. Thus, some say we are years, decades, away from getting near to the stage at which robots can think for themselves at a level at which they become a threat, and even then it may well never happen. This seems far too complacent, and whether it is years, decades or centuries away, we should care about, and plan for, the possible adverse impacts on humanity.

Some economists have also been too complacent in arguing that unemployment will not be a problem, as robots will open the way to a host of new jobs that 'we can scarcely imagine'. Relying on the assumption that major innovations in the past have always resulted, after possible initial problems, in no reduction in employment, misses the point that this technology is different. Other technologies augmented human capital. This one replaces human capital, and the past may no longer be a good guide to the future. The critical question is the proportion of human activity robots replace, and on that we just do not know. Other economists do not fear unemployment per se, but they do fear increased inequality and point the way to solutions. Still other economists are arguing that, yes, unemployment is also going to be a problem.

Robots replace humans as they are potentially superior on all dimensions. In terms of manoeuvrability, they will be able to change from wheels to legs, go underwater and then into the sky. They are not limited to two hands, they can have all-round telescopic and x-ray vision, and they can detect sounds humans cannot as well as other facets of the environment such as levels of radiation. They can calculate many times faster than humans, and instantaneously interact with other robots. They are still a long way behind humans in many areas such as empathy, creative thinking and even something as simple as distinguishing an apple from a pear. But these problems are likely to be solved. If we were put on this Earth as part of an experiment to see whether starting from scratch we could construct machines which outmatch us and other life forms in every respect, we are getting close to the 'successful' end of the experiment. The question is: what happens next?

There are two sets of problems that might arise from the robot revolution. The first are fairly direct: robots replacing jobs, robots spying on people in their home, robot criminals and hacking into robot systems to cause massive chaos. But the second are the unintended consequences, which may take centuries to play out. Some we can foresee, others we cannot. For example, in relying too much on robots to do our thinking, our minds may regress. In relying on robots in the home, people may have less contact with other people, may have fewer children, and may marry less. Of course, the opposite may happen: with increased leisure time, people may socialise more.

We do not know. But we should be prepared. We should also try to implement policies that pave the way for the positive impacts.

In terms of what to do about robots, countries in isolation have no choice: they must set out to take full advantage of robots in the workplace, hopefully in forms which complement human labour rather than replace it, and which enhance wellbeing and do not reduce it. If they do not do this then they fall behind other countries and are then likely to suffer more, not less, unemployment. They must also regulate robots, possibly licensing some. Redistributory policies will help with the inequality problems. Other issues, such as robot soldiers, are best solved, if at all, by international agreement.

These are all issues we deal with in this book. It is written from the perspective of an economist, but one convinced that we must have some understanding of the underlying science. Hence there is a chapter on this, and a related chapter looking at all the different types of robots. Another chapter reviews the history of robots to the present day, both in fact and in fiction. There are two empirical-based chapters, where we examine the impact of robots on the labour market and also people's attitudes to robots. Two more chapters look at the impact of robots, positive and negative, on the economy, society and people and what policies to pursue in order to maximise the benefits and minimise the downside. Finally, there are two chapters on innovation and what lessons we can draw from our analysis of robots for theories of innovation. Thus we argue that such theories are too focused on the process of innovation and not enough on its impact on people and scociety.

This does seem to me to be one of the most important issues facing us today, a perception that grew as I wrote the book. Governments must pursue policies that ensure that we take full advantage of the positive potential of robots and to make their economies as productive as possible. But they must also, either acting on their own or collectively, try to control some of the negative impacts of robots. I must admit that, whilst recognising the benefits robots will bring, I do have substantial concerns about the impact robots will have on people, our society, our economy and humans per se. Whether it is possible to prevent all, or even just the worst of, the negative impacts I do not know. I am not overly optimistic, but I also know that it is critical that the attempt must be made.

In a sense the book is telling a story, well before the end of even the first act. I am fairly sure that part, at least, of the second act will take us by surprise, and we will be even more surprised by the end of the final act. But as economists and social scientists we cannot wait till the end of the story before taking a view. We have to make policy decisions now, both to maximise the benefits and to try to protect society against the problems.

A few final thoughts. Robots will change just about everything, sometimes because of what they do and sometimes because of how people and governments react to them. So as individuals make the best of the opportunities, this will open up. Your world will change beyond recognition over the next 30 years and in many ways you will benefit enormously. Finally from my own perspective, robots will have a fundamental impact upon economics as well as the economy, and it will be fascinating to see this unfold.

Acronyms

AI	artificial intelligence
AIC	Artificial Intelligence Center (USA)
AGV	automated guided vehicle
AMR	autonomous mobile robot
ANN	artificial neural network
AV	autonomous vehicle
BCI	brain computer interaction
CAM	computer assisted manufacturing
CBD	Convention on Biological Diversity
CPU	central processing unit
CTBT	Comprehensive Test Ban Treaty
DARPA	Defense Advanced Research Projects Agency (USA)
DFID	Department for International Development (UK)
DOF	degrees of freedom
EEG	electroencephalogram
EU	European Union
GM	genetically modified
GPT	general purpose technology
GPU	graphics processing unit
IC	integrated circuit
IEC	International Electrotechnical Commission
IEEE	Institute of Electrical and Electronics Engineers
IFR	International Federation of Robotics
IPR	intellectual property rights
ISO	International Organization for Standardization
IT	information technology
LAR	lethal autonomous robot
LAWS	lethal autonomous weapons systems
ML	machine learning
OECD	Organisation for Economic Co-operation and Development
R&D	research and development

SLAM	simultaneous localisation and mapping
SRI	Stanford Research Institute (USA)
T&R	teach and repeat
UAV	unmanned aerial vehicle (or drone)
UCL	University College London (UK)
UGV	unmanned ground vehicle
UK	United Kingdom
UNSC	United Nations Security Council
USA	United States of America

1. Innovation

1.1 INTRODUCTION

Robots and automation figure at the end of a long line of innovations. In the past, innovation has, by and large, been positive for the world. There have been some problems, some as the result of side effects as with the steam and petrol engines and subsequent carbon dioxide emissions. Other problems, such as with nuclear power, are due to man's ability to use innovation for both good and evil. However, innovation has facilitated an ever-increasing standard of living, not just in providing more and better quality goods, but new goods and new opportunities. Hence today we can communicate with people on the other side of the world visually as well as with mobile phones. We can then travel to the other side of the world in a few hours, or if we prefer watch a football match in another country from the comfort of our armchairs.

Innovation has been with humans from the very beginnings of their time on Earth. The use of fire to cook and to warm, the spear and the bow and arrow, agriculture, the plough, bronze, iron and the wheel, all had major impacts on human development, just as innovation continues to do so today. It has also been true that innovation has changed societies. The onset of farming fundamentally changed society, and, somewhat later, the printing press facilitated the reformation. But if innovation has been with us since our early beginnings on this planet, its pace is quickening. Today we seem to get major innovations if not every decade then at least every two decades. It is also noticeable how innovations are often linked. It is relatively seldom for a major innovation to appear without it having built substantially on the work of others. For example, the first commercially successful helicopter was designed in 1939 by Sikorsky. However the first piloted helicopter had been developed some 32 years earlier by Paul Cornu. But the concept can be traced all the way back to Da Vinci – a link that can also be made with robots. It is also apparent how innovation has tended to be focused on specific countries at different times. Hence in the period 1700–1950 many of the world's major innovations were born in the United Kingdom (UK). Since 1950, the USA, and in particular the West Coast, has been critical.

Some innovations have a fixed life cycle and are then replaced, as the word processor replaced the typewriter. Some are reborn under new, more advanced guises, as with the windmill, which today is a significant generator of electrical power. Initially most innovations were process innovations – new and more efficient ways of doing things. But the average man in 1500 was living pretty much as the average man in 200. The major difference was one of quality and quantity facilitated by innovation, but the basic consumer goods of food, drink and clothing were essentially the same. The way food was cooked was also essentially the same. Sea travel had progressed considerably with navigational aids, but wind was still its energy source. Land travel still essentially involved the horse. Some limited progress had been made on the medical front and man was more efficient at warfare. But, in the late eighteenth century we begin to get more of a change. The railway was a revolution in transport, a fundamental innovation that had enormous spillover effects. Equally, somewhat later, the telegraph revolutionised communication.

Then, around the beginning of the twentieth century, another revolution took place. Innovation was changing people's lives not by supplying more and better quality of what had been their lot for hundreds of years, but by changing that lot. The vacuum cleaner was invented in 1901, the first of the household goods that would transform women's role in society. A few years earlier the first motorised vehicles appeared and these would provide freedom of movement, at first to a privileged few such as Mr Toad, but then to the masses. This freedom was eventually translated to the air, courtesy not just of the Wright Brothers but also the low cost airlines. Now, at the beginning of the twenty-first century, we have become used to the idea that our lives will constantly undergo revolution, that just as 30 years ago we wrote letters, searched for telephone boxes and used paper maps to find our way around, so in 30 years' time things we do now will become revolutionised.

1.2   THE PROCESS OF INNOVATION

In a sense, Schumpeter is to the supply side of economics what Keynes is to the demand side. Both economists were born in the same year, both were giants of their and any age. Initially Keynes' work had the greater impact, certainly from 1940 onwards, but Schumpeter's work endures, still inspiring economists and setting the tone for much of today's research agenda. Schumpeter's view of innovation was that a discovery could be made at a certain time, but it was unlikely to be developed by entrepreneurs until old technologies had reached the end of their life and firms were having difficulty making profits. At this time entrepreneurs would be receptive to new ideas.

Major or blockbuster innovations are innovations of the first order that either provide a revolutionary new product or a revolutionary new way of making existing products. In their wake, more minor innovations follow. The new technology opens up new possibilities. Because they are new, there are no existing firms in the field and it is often the small-scale entrepreneur who seizes the opportunity, although they often do not stay small. Many entrepreneurs come in at the second stage of innovation, for example with the railways, the hoteliers, and in the wake of artificial intelligence and the Web came Facebook and Google. In general, process innovations often pose a threat to jobs. Product innovations may also pose such a threat, for example the railway replaced the canals and the stage-coach, but they also create new ones and often there is no product being replaced: the innovated product is entirely new. Eventually this new wave of innovations will have reached its full potential and once more the process begins. Thus the entrepreneur is at the heart of innovation. Sometimes they are the inventor who brings their product to the market place. Sometimes they make use of someone else's research and bring that to market. Schumpeter at different times thought innovation was best served by small firms, as with Steve Jobs working in his Californian garage, or by large firms, such as Apple, who have substantial resources to throw at innovation.

A general purpose technology or GPT is a particular form of blockbuster technology. Specifically, it is an innovative method of producing, important enough to have a substantial and prolonged aggregate impact. It has also been defined as a technology that becomes pervasive, improves over time, and generates complementary innovation (Bresnahan and Trajtenberg, 1995). Sachs (2018)

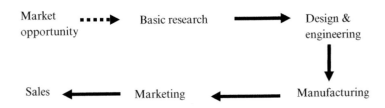

Figure 1.1   Linear innovation

argues that GPTs result in deep structural changes, raise GDP, disrupt production processes and restructure labour markets. They impact on the distribution of income and wealth and change human geography and demography. Both steam and electricity would be regarded as GPTs.

Until the 1990s, the linear model of innovation policy was dominant. This viewed technical change as happening in a linear fashion from invention to innovation to diffusion. It led to an emphasis on research and development (R&D) infrastructure provision, financial innovation support for companies, and technology transfer. Variants of the linear model include technology push and market pull models. The stages of the former in the original linear model are shown in Figure 1.1. Basic science is the start of the process, followed by the design and engineering, manufacturing, marketing and sales stages. However, this model tended to ignore prices and other changes in economic conditions that affect the profitability of innovations. Thus, the market pull model starts with the identification of a market need first, followed by technology and product development, production, and marketing. The sequence seems similar, but the implications are different. The first presupposes that the basic impetus for new development comes from the scientist, the second that it comes from market forces. It presupposes that innovation can be 'made to order', that once a gap has been identified someone will fill that gap with a suitable technology. Sometimes this is the case; sometimes less so. In reality, both forms of innovation are possible in the real world and the scientist can develop a new technology that the entrepreneur sees as filling a gap – in some cases a previously unrealised gap.

Innovation can also come about through a change in the environment that forces innovation on the private sector, which may happen without any external policy intervention. The case of the sweet potato in Uganda illustrates this (Hall and Clark, 2010). The onset of disease forced a shift from cassava to the sweet potato not only as a source of food but also as a cash crop. Farmers experimented with, and introduced, new varieties of sweet potato and also developed new products for the market. Much of this they did on their own initiative with little outside help. It required changes not just in economic production, but in diet and culture, and without the necessity of survival forcing change it would not have happened. Policy still has an important potential, if often unrealised, role in: (1) supplementing the limited knowledge and resources of, in this case, farmers; and (2) shifting the boundaries and constraints that limit endogenous innovation. For example in this case, the research institutes should have been immediately focused on this problem and in dialogue with the farmers. They were not. It also illustrates something that is often forgotten in the literature: that innovation is a global phenomenon and has no less a part to play in developing countries than in richer ones.

Since the turn of the twenty-first century a new understanding of the nature of the innovation process has emerged, which stresses its systemic and interactive character (Todtling and Trippl, 2005). The emphasis on feedbacks, interactions and networks was incompatible with the linear model where the flow is uni-directional (Freeman and Louca, 2001). This approach offers a less deterministic version of the technology push argument, while still emphasising the role of science and technology. It also stresses that the first stage can come from either internally generated knowledge or knowledge acquired from outside the firm. The innovation systems approach argues that innovation should be seen as an evolutionary, non-linear and interactive process, requiring intensive communication and collaboration within firms and between firms and universities, innovation centres, financial institutions, standards setting bodies, industry associations and government agencies. End users too can provide vital feedback. This approach has led to a realisation that there should be a shift away from the traditional firm-oriented perspective towards a more holistic view and one that emphasises inter-organisational arrangements, that is, a move towards a more system-centred approach of innovation policy (Nauwelaers and Wintjes, 2003). This is particularly the case when the innovation is combining science from different fields such as robotics and nanotechnology, or even robotics itself. This does not mean that focusing on R&D and on the technological aspects of innovation is the wrong policy, but that it needs to be complemented with the other aspects of innovation such as organisational, financial, skill and commercial. The Triple Helix model of innovation brings university, industry and governments together (Etzkowitz and Leydesdorff, 1995). One of the underlying ideas is that in a knowledge-based economy, the potential for innovation and economic development lies in a prominent role for the university. It also allows for the joining of agents from university, industry and government to generate new institutional forms for the production, transfer and application of knowledge. Hence the Triple Helix incorporates a wide diversity of approaches.

1.3   THE STAGES OF INNOVATION

There is relatively little disagreement as to the building blocks of innovation as expressed in Figure 1.1, although some might add more blocks. But the disagreement tends to focus on how they are related and inter-related. Thus, in this section we will examine these building blocks and then turn to the relationships between them. The building block over which there is probably most disagreement, or if not disagreement then variation in perspectives, is that relating to basic science.

1.3.1   Basic Science: Discovery

Invention can arise from: (1) scientific curiosity and ego; (2) need; and (3) a search for profits. The discovery of Vecro provides an example of the first of these. The Swiss engineer, George de Mestral, found seed pods sticking to his socks. Motivated by curiosity, he discovered the pods had small hooks that had caught themselves into the wool of the socks. He reproduced these hooks in woven nylon as a way of fastening clothing together instead of buttons and zips. One thing led to another, and this included joining the chambers of an artificial heart and securing objects in space with zero gravity. 'Need-driven innovation' is where there is a specific problem that needs solving and scientific resources are devoted to solving that problem. 'Curiosity-driven innovation' often happens by accident. As such it is more difficult to model than need-driven invention. An example of profit-driven invention is when a firm has a research arm, which is given the specific instructions of finding new discoveries with the potential to generate profits.

A combination of need-driven and 'accidental invention' is the 'unintended consequences invention'. A specific invention opens up new possibilities. The original invention was not driven by a need to open up these new possibilities – they happened. The need can reflect a problem that needs solving or a cheaper and more efficient way of doing something. Once the need has been identified, resources are devoted to discovering a solution. It is not simply that the resources are devoted at this objective and a solution found in an almost mechanistic way. Sometimes this is the case, but often a feat of imagination or some blinding insight is a critical part of the process. But this insight would not have happened unless the focus of attention had not been devoted to the problem. Of course, this insight is often a rough diamond and still needs considerable work before it can be presented as a working solution to the original problem.

An example is mobile finance. This possibility was opened up by the mobile phone. This is illustrated with respect to Kenya. Mobile phones have revolutionised telecommunications in Kenya and from a very low base. At the end of the fourth quarter of the 2016–17 financial year, mobile phone coverage was 88.7 per cent of the population. There were 28.0 million active mobile money subscriptions and a total of 480.5 million transactions amounting to 96.4 billion US$ in this fourth quarter. In addition, goods and services purchased over mobile platforms amounted to 6.6 billion US$. Person-to-person transfers were valued at 5.2 billion US$.[1] This has transformed the Kenyan economy, facilitated the growth of small businesses, increased the transfer of funds to rural communities and extended financial inclusion to millions. At the heart of this revolution is the M-Pesa system for money transfer and financial services. This permits users to swap cash for 'e-float' on their phones, which they can then send to other mobile phone users, who can exchange the e-float back into cash. This developed from research funded by the UK's Department for International Development (DFID), who observed that Kenyans were transferring mobile airtime as a substitute for money. Together with a UK private mobile service provider they filled this gap. Hence this is an example of one innovation (the mobile phone) leading to another (mobile financial services). It is

---

[1] All these data from https://ca.go.ke/wp-content/uploads/2018/02/Sector-Stat istics-Report-Q4-2016-17.pdf.

also an example of how market opportunity, government involvement and private sector firms are often interconnected in innovation.

The identification of need at the corporate level can be a widely
recognised need on which other firms are also working, as with new
drugs to meet specific diseases, or it can be something unique to the
firm itself. In the latter case the identification of need is a critical part
of the innovation process and is the first stage where imagination
plays a critical role. Examples include the invention of cat's eyes by
Percy Shaw in 1933 to help with the problem of unlit roads following the advent of the car. In this case the realisation of need and the
outline of a potential solution came at the same time. Shaw had been
using the polished strips of steel tramlines to navigate, something he
realised when the tram lines were pulled up. Hence potential solution
and need appeared at one and the same time. The exact process by
which intuition takes place and insights are formed is more in the
realm of psychology than economics. One example is that of bisocia-
tion (Koestler, 1964) a process by which two apparently unconnected
ideas are linked.

1.3.2 Development

The 'valley of death' is the stage between a new invention or discovery being translated into a product or service. The company needs to
find sufficient money to develop the prototype until it can generate
sufficient cash, through sales to customers, which would allow it to be
self-sufficient and grow. Many do not cross the valley, hence its name.
One reason for this failure is linked with an inability to attract sufficient finance to develop the project. Another feature of the valley is
that often when crossed, we find ourselves in another country from
where the research was done, with the finance coming from foreign
firms. Raising finance can be difficult due to enhanced uncertainties.
Innovations are, almost by definition, riskier than other investments,
although offering the prospect of greater returns. This is emphasised
by the World Bank (2010), who observe that most innovations fail. In
order to bring the research to market, translational or developmen-
tal research is frequently needed. Even with successful innovation,
feedback from manufacturers or from retailers and consumers may
modify the innovation. This can often lead to unexpected ideas and
products, but it can also lead to increased expense in developing
the technology. In general, governments cannot directly help in the

raising of finance for the development, as opposed to the research stage. But they can act to bring together sources of finance and firms seeking to develop new products.

1.3.3   The Process of Technological and Innovation Diffusion

Rogers' (1995) theory of technology adoption has been highly influential. This focuses on the user of the innovation, for example the consumer. There are five stages to innovation adoption. Firstly is where the individual becomes aware of an innovation. Secondly, the individual obtains enough information on the innovation to make a judgement on its usefulness. In the third stage the individual chooses whether to adopt. Stage four sees the individual buying the innovative product and, finally, in stage five, the individual re-evaluates whether to continue with the innovation. Rogers also identifies five innovation characteristics that influence adoption. The first relates to the advantage of the innovation over alternatives. Secondly, compatibility may be negatively correlated with this, and relates to the similarity of the innovation to previous ways of doing things. Thirdly, the greater the similarity, the less complex it will be to understand the innovation. Fourthly, trialability is when the individual can experiment with the innovation prior to adoption. Finally, observability is linked to how many others are using the innovation – which, with some innovations such as the mobile phone, relates to relative advantage.

Innovation adoption is often characterised by an S-shaped adoption curve, with there being a relatively small number of initial adopters, following which, if the adoption is successful, adoption accelerates. Early adopters tend to have high socioeconomic status, to be literate and relatively intelligent with good methods of communication (Straub, 2009). Innovation and knowledge tends to be first diffused in large towns and cities (Henderson, 2007). It then becomes diffused to other areas (OECD, 2010). Thus, in general, stimulating innovation in a new country is in most cases a two-stage process. Firstly, it needs to be targeted at the large cities, where it will be most successful with ready acceptance (ibid.). Secondly, it then needs to be diffused to the rest of the country. If not, innovation will simply widen inequalities within the country, as can be observed, for example, in India and China. Consistent with this, the World Bank (2008) repeatedly emphasises that technology diffusion differs within countries as well as between them. This discussion largely relates to product innovation. Process innovation relates more to firms, although similar considerations apply.

Innovation involves new knowledge, but this does not necessarily come about via R&D within a firm. Instead the firm can acquire this knowledge from other sources. The knowledge may be entirely new, or new to market (that is, new to a country, region or town). When the latter is the case, one of the first stages of innovation involves knowledge diffusion, together with knowledge adaption into a form suitable for 'market X', again the country, region or town. The extent of adaption will depend upon the knowledge itself and its application in another market that most closely resembles X. There is a whole continuum of innovations from very large ones to much smaller ones. A key barrier to knowledge diffusion is distance (Jaffe et al., 1993). Geographic distance tends to increase the costs of transferring knowledge and technology, by for example reducing knowledge spillovers. Distance is related but not restricted to spatial distance and Rosenthal and Strange (2004) also emphasise both social or economic distance. Information technology (IT) may also be changing the relationship between knowledge diffusion and spatial distance.

1.4   OPPOSITION TO AUTOMATION AND NEW TECHNOLOGY

The example everyone knows is of course that of the Luddites. But even prior to the Luddites there were concerns that automation would put people out of work. One early example is that of William Lee's stocking frame knitting machine developed in England in 1589. Queen Elizabeth I refused him a patent, saying: 'Thou aimest high, Master Lee. Consider thou what the invention could do to my poor subjects. It would assuredly bring to them ruin by depriving them of employment, thus making them beggars' (cited in Acemoglu and Robinson, 2012, p. 182). Apart from anything else, this illustrates how much easier it is to gain a patent today. Shortly afterwards, William Lee had to leave Britain. Frey and Osborne (2017) argue that the Queen's concerns were linked to the hosiers' guild's fear that Lee's invention would make obsolete their artisan members' skills. The guilds systematically opposed technological progress and new inventions where this would be against their members' interests (Kellenbenz, 1974). However, by the mid-seventeenth century, in contrast to Continental Europe, the craft guild in Britain had declined in importance and influence (Frey and Osborne, 2017). This helped pave the way for a change in attitudes reflected by legislation passed in 1769 making the destruction of machinery punishable by death (Mokyr, 1990). The decline of the gilds helped pave the way for the industrial revolution. Their opposition to much innovation was from their own perspective partially justified. Most of the technologies of industrial evolution substituted unskilled or semi-skilled workers for skilled artisans, a process that accelerated with the development of steam power (Goldin and Sokoloff, 1982). Hence, in the nineteenth century, new technologies tended to substitute for skilled labour through task simplification (Goldin and Katz, 1998), Thus as with the current technological revolution, innovation shifted the balance between skilled and unskilled workers as well as between labour and capital.

During the nineteenth century, establishments grew in size, as improvements in steam and water power technologies facilitated the adoption of powered machinery that resulted in large productivity gains from combining higher capital intensity with greater division of labour (Atack et al., 2008). However, in the twentieth century, the balance shifted back in favour of the skilled worker with the switch to electric power and away from steam and water power. This shift, together with the development of continuous process and batch production methods, reduced the demand for unskilled manual workers and increased the demand for skilled workers (Goldin and Katz, 1998). The increase in production and market size also increased the number of managerial tasks, which in turn required more white-collar, non-production workers (ibid.). Hence technological change has continually changed the relative advantage of skilled versus unskilled labour over long waves lasting many decades.

1.5 INNOVATION CASE STUDIES

In this section we consider several specific innovations that help us to understand the process of innovation.

1.5.1 Case Study 1: The Spinning Jenny

The invention of the spinning jenny is analysed by Allen (2009). It was invented by James Hargreaves in 1764 and was, to an extent, a market-driven innovation. Cotton production had difficulties meeting the demand from the textile industry. However, Allen argues that its introduction in England was a consequence of the high wages there, relative to those in both India and Europe. Without techno-logical innovation it would have been difficult for English industry to compete, and viewed in this respect, this and other developments preserved jobs and incomes in England, despite all the concerns amongst, and the opposition of, the ordinary spinners. Thus, to an extent it was the cotton industry and workers outside the UK who were disadvantaged by this technology more than British workers. This is seldom a concern of research: when it looks on the employ-ment consequences, the focus is generally within the same country. This is a mistake.

Prior to the spinning jenny, spinners used the spinning wheel. The cotton had previously been cleaned and carded, which is similar to being combed, to produce a loose strand of cotton called a roving. The spindle was in front of the spinner and the wheel on the spin-ner's right. One end of the roving was attached to the spindle and the spinner held the rest in their left hand. Drawing their left hand away from the spindle lengthened and thinned the roving, which was then twisted. This twisting made the yarn strong. Hargreaves' key objective was to allow the spinner to operate multiple spindles rather than one. He had tried operating several wheels by holding all of the threads from each in one hand, but that proved impossible with horizontal spindles. The key insight was to use vertical spindles with a single wheel. The earliest jennies had 12 such spindles, but soon 24 became the standard. The increase in productivity was enormous, but it came at a cost, with a single 24-spindle model costing 70 times as much as the spinning wheel. But they were still small enough to be operated in a spinner's cottage.

This resulted in a reduction in the price of yarn, which angered the spinning workers in Lancashire where Hargreaves lived. They broke into his house and smashed his machines. He then moved to Nottingham where he again began the production of jennies. He patented his invention in 1770, and it continued in common use until superseded by the spinning mule in about 1810. Hargreaves was an illiterate weaver and carpenter who perceived there was a problem and produced a solution. It took him several years to perfect the invention and in this he was financed by a small-scale farmer. His invention mimicked, on a greater scale, the actions of a spinner and wheel and, in Allen's (2009) words, was 'not rocket science'. Nonetheless its impact was enormous. One final point to note: this was an invention that could have been made many decades earlier.

1.5.2 Case Study 2: The Railway

The key technology was the steam engine developed by James Watt. This of course was, in its own right, a key element of the industrial revolution, but when developed by Richard Trevithick who used high-pressure steam to increase the power–weight ratio, it paved the way for a steam engine to power its own movement. An early engine was in use by 1804, running on rails at a Welsh ironworks. The first proper steam locomotive pulling wagons was the Stockton and Darlington line, opened in 1825, which reached a dizzying, for the time, 15 mph. George Stephenson's engine was, however, unreliable and it was not until the completion of the Liverpool and Manchester line that the railway age truly began. Initially the railways were built with a view to hauling goods and raw materials. From the earliest days, too, carrying the mail was envisaged. What was less expected was the substantial demand from the public to travel. From this time on, the railways prospered, but were hampered by the fact that the gauge, the distance between the rails, was not standardised, thus limiting continuous travel in the same train. This was resolved by a Royal Commission on Railway Gauges in 1845 that had to choose between the broad gauge of Brunel's Great Western Railway and the narrow gauge used by many others. The Commission came down on the side of the latter, with a width of 1435 mm. This became law in 1846. This standard has since been adopted by much of the rest of the world. Further government intervention, via Gladstone's 1844 Railway Act, forced the companies to provide at least one daily train costing no more than a penny per mile, thus paving the way for transport for all and not just the richer people.

The railways did put people out of business and led to job losses particularly on the canals and in the stagecoach industry, including the many stagecoach inns in more remote areas. In 1830 the industry employed in excess of 30 000 people and there were more than 1000 turnpike companies who maintained the roads (Wolmar, 2009). Both were in decline from 1840 onwards. Towns not served by the railway also went into decline. But by 1860 more than 300 000 people were directly employed on the railways (ibid.) and there were many employed as a consequence of the railway. Railways facilitated developments as varied as the growth of the whisky industry and professional football. The impact on lifestyles was substantial – not least, the ability to transport food rapidly to large towns improving the average person's diet. As with any new technology there tended to be various fears about their impact. Thus, there were concerns that cows would stop producing milk because of the noise and that the smoke would turn sheep black. Looking back today, many of the early concerns seem laughable. But few people came up with the even more far-fetched concept that the railways and the steam engine would contribute to carbon emissions that would eventually lead to climate change.

1.5.3   Case Study 3: The Integrated Circuit (IC)

The transistor acts like a switch. It can turn electricity on or off. Its use in early computers substantially reduced their size. But they were unreliable. The solution to this problem came in 1959, when Robert Noyce of the Fairchild Semiconductor Corporation patented a silicon-based IC and Jack Kilby of Texas Instruments patented miniaturised electronic circuits. In many respects the revolutionary nature of the IC lay with making everything out of one material, mainly single crystal silicon wafers, including both transistors and the wiring, although the idea had been around for some time. Prior to this, these elements were used to make transistors, but carbon was used to make the resistors and the connecting wires were made of copper.

There were two other problems facing the IC. Firstly, there was no way to electrically isolate the different components on a single semi-conductor crystal. A semiconductor has an electrical conductivity between that of a conductor such as copper and an insulator such as glass. Their conducting properties may be altered by doping, that is, the controlled introduction of impurities into the crystal structure. A p–n junction is a junction between two differently doped regions. The isolation problem was solved in 1958 by Kurt Lehovec, who worked at the Sprague Electric Company using the p–n junction concept. It substantially reduces the electrical flow and hence any degree of insulation can be achieved by having several p–n junctions in series. Sprague was not interested in the idea and Lehovec filed his own patent application and then left the USA for two years. The second problem was that the only effective way to create electrical connections between the different components of an IC involved using gold wiring, which was very expensive. This was solved by Robert Noyce, building on Jean Hoerni's planar process, which created a flat surface structure protected with an insulating silicon dioxide layer. This views a circuit in two dimensions (a plane) and allows the use of photographic processing, making a series of exposures, thus creating silicon oxide (insulators) or doped regions (conductors). Noyce connected the transistors on the wafer with aluminium wires placed on top. Both Noyce and Hoerni were among a group of eight who had founded the Fairchild Semiconductor Corporation, where this work was done.

In 1960 a group, under the leadership of Jay Last, started by Noyce at Fairchild, produced the first planar IC. However, it was Texas Instruments who received contracts for planar ICs for space satellites and ballistic missiles. The ICs for the onboard computers of the Apollo spacecraft were, however, designed by Fairchild. As a consequence of this demand and the existence of economies of scale, the price of ICs dropped enormously, and hence, as is often the case, we see defence-related government agencies playing a leading role in the development of an innovation that would eventually have huge private sector demand.

The microchip of today is vastly different from earlier ones and one fingernail-sized chip can contain ICs with millions and even billions of components, hence facilitating the substantial increase in power of both computers and mobile phones. But it is still based on the work of these early pioneers. In passing, we note that Kilby got the Nobel Prize for his work in 2000. Noyce died in 1990 and hence could not be nominated. There is, however, a perception that the contributions of others have been somewhat undervalued (Lojek, 2007, p. 156).

1.6 REFLECTIONS ON INNOVATION

The spur for innovation is some combination of a desire for monetary gain and scientific curiosity. The innovation may stem from the work of the scientist, today often in a university environment, in the past less so, driven by a desire to understand the world, to make discoveries about that world and make it better, and driven too by an ego, with the scientist seeking recognition from their peers and perhaps the history books. But the innovation may also stem from a perception of a market need that it would be potentially profitable to fill. This perception may come from an established firm, as with Texas Instruments, or from the individual working alone, as with James Hargreaves and the spinning jenny. In the latter case, the individual often seeks to market the device themselves, frequently setting up a new firm in the process. The individual is then both entrepreneur and inventor, which may be difficult as they are combining two distinct skill sets.

The invention, once successfully marketed, impacts on society, generally increasing wellbeing but sometimes with adverse side effects. Railways were a disruptive technology that destroyed thousands of jobs, putting many people out of business. The same is true for the spinning jenny. But over time many more jobs were created, both in the railway industry and in other industries. They are clear examples of what Schumpeter termed 'creative destruction'. Railways also fundamentally changed society and the economy. Some of these benefits were seen from the moment the innovation became available and were in part the initial spur for the innovation. But some were not foreseen. The growth of passenger transport took the early railway innovators somewhat by surprise, but they were then quick to exploit the opportunities. This is another feature of innovation: the unintended consequences. In this sense, when you buy a ticket for innovation you buy a ticket into the unknown, and these unintended consequences may take decades to play out. This is one rationale for governments to play a part in the innovative process, as in the helix models, but not to spur innovation, rather to control it and its side effects and also to maximise the benefits to society. An example of the latter would be the insistence on the penny per mile railway ticket. Hence governments play a role not just in correcting for market imperfections, as with specifying the standard rail gauge, but in protecting society and people from the more harmful effects of innovation. Governments also lay out the ground rules for innovation, with the patent laws that specify for how long the firm with the patent can uniquely benefit from that innovation.

1.7   IS INNOVATION GOOD FOR SOCIETY?

Up until now the answer has largely been yes, although a qualified yes. Few would wish to go back even 100 years and have virtually no domestic appliances and no ability to communicate with others over long distances. But that does not mean that all innovation is good and there is not some innovation that in hindsight we would wish to have stopped, and other innovation we would have done differently. Have developments in nuclear physics, for example, on balance been beneficial to society? More generally, Stiglitz (2018) argues that the free uncontrolled market may result in innovation that leads to greater unemployment and inequality than the socially desirable levels and thus reduce societal wellbeing below what it could be. Indeed, the result may not even be output maximising.

The problem is that innovation is driven by scientific curiosity and greed, the desire to make profits or of governments to enhance their defensive and security capabilities. If it is possible to make something and it will make a profit then it will be made, unless governments, all governments, make it illegal to do so. The main driver of innovation is often not to make society better – that is a side effect. There are exceptions to this of course, as with the example of mobile finance in Kenya. But frequently it is a side effect: an innovation makes profits because people or governments are willing to pay for the innovation, because it makes the former's lives better or improves the latter's ability to provide public goods. But this does not automatically mean it makes society as a whole better off. Hence, although until now innovation has largely been beneficial to humankind, this is not something we can bank on continuing into the future. There is no fundamental dynamic to innovation that guarantees it will always improve our lives and our society. Nor has innovation in the past been implemented optimally. There is a balance to be had between the disruptive and beneficial effects of a new technology, and it is not obvious that the invisible hand always achieves the right balance. The invisible hand is good at providing the incentives for innovation. It is less successful at protecting the losers or controlling the speed of innovation. That must also be a role for government.

Because there is no guarantee that innovation always benefits society, and because a desire to do good is only part of the incentive for innovation, there is a rationale for regulating scientific endeavour on several dimensions. Firstly, we imagine a world where some research may be allowed and other research not allowed. In actual fact, this is the world we sometimes at least aspire to, as with the case of human cloning (Langlois, 2017). The difficulty lies in getting all countries to agree to this and then to enforce it – a difficulty that may be possible to overcome. Secondly, research and innovation that are allowed may still be restricted and subject to regulation, to maximise the benefit and minimise the risk. This is increasingly being seen, in part because governments are becoming more involved with stimulating and directing innovation. Apart from regulation, there is also a case for increasing the emphasis on the teaching of ethics to scientists both whilst students and subsequently (Reiss, 1999).

2. The history and development of robots

2.1 INTRODUCTION

According to ISO 8373, an ISO standard, a robot is defined as 'an automatically controlled, reprogrammable, multipurpose manipulator with three or more axes'.[1] Some definitions refer to it looking like a human. To an extent this is similar to the dictionary definition of a machine controlled by a computer that is used to perform jobs automatically.[2] There are three main elements to robots: manoeuvrability, a sense of space or location, and a guiding intelligence. Manoeuvrability includes both the ability to physically change location and to move 'arms' and use them to grasp objects. It is important to bear in mind that robots are machines. At least at this point in time, they cannot think as humans do, nor can they see or hear. All of this has to be done artificially with the sensors doing 'the seeing' and 'the hearing', and the processing unit doing the 'thinking'. When you speak to a humanoid robot, it is not, often despite appearances, reacting to you emotionally. It is searching through its memory banks for the appropriate response. It is not happy when it makes the right response, nor embarrassed when it makes a social gaffe, although it may learn from the experience.

Robots differ from other forms of automation such as computer assisted manufacturing (CAM), which is the use of software to control machine tools. This differs from robots in the sense that the machine has not fundamentally changed from a manually operated one. Until recently the application of robotics was mostly limited to recurring mechanical actions or actions coordinated by a human. As a result of recent advancements in the science, there are more and more actions that can be properly performed by robots. Moreover,

---

[1] https://www.iso.org/standard/55890.html.
[2] http://dictionary.cambridge.org/dictionary/english/robot.

due to the application of artificial intelligence (AI) and machine learning (ML), robots are becoming more autonomous and human intervention during a robot's operation is declining. There are also more examples of robots doing things that humans simply cannot do. For example, in paediatric surgery robots offer the advantage of stereoscopic vision and magnified view (Cundy et al., 2014).

2.2 EARLY ROBOTS

In this chapter we review the development of robots up till the present. Robotic devices can be traced back to ancient times, although in the past they largely focused on manoeuvrability rather than a sense of location or a guiding intelligence. Hence, in the Middle Ages, Leonardo da Vinci built a robotic knight, that is, a knight suit with gears and wheels that was capable of sitting down, standing up and moving its head.[3] Ismail Al-Jazari, who designed several robots, including a four-musician robot band, in twelfth-century Turkey, had influenced Da Vinci. Despite their limited nature, these early robots have informed future developments – thus the da Vinci robot as a basis for robot design for NASA. From very early times Japan has played a role. Hisashige Tanaka published the influential *Karakuri Zui* (*Illustrated Machinery*) in 1796 and built complex toys, which for example could serve tea or fire arrows drawn from a quiver (Hornyak, 2006). Later examples, of which there are many, include a robotic soldier with a trumpet, built in Dresden in 1810 by Friedrich Kaufmann,[4] and Elektro, a humanoid robot shown at the 1939 World Fair in New York.[5] Elektro could perform several tricks including walking, speaking and blowing up balloons. One of Japan's first robots was Gakutensoku, built in 1929 by biologist Makoto Nishimura.[6] All of these devices were mechanical and could not 'think' or get anywhere near doing so. Hence, they anticipated the mechanical aspects of robots, in particular movement, but not the intelligence aspect. They were built before the age of computer-

---

[3] http://www.da-vinci-inventions.com/robotic-knight.aspx.
[4] http://www.deutsches-museum.de/en/exhibitions/communication/computers/automata/.
[5] https://www.youtube.com/watch?v=AuyTRbj8QSA.
[6] http://jpninfo.com/39826.

controlled servo-mechanisms, that is, mechanisms used to automatically correct performance via an error-sensing feedback. For this, of course, we needed developments in computing as well as sensors.

To an extent, robotisation is simply another step down the line of automation. The dictionary definition of automation, a term that appeared in 1936, being coined by Delmar Harder (Hitomi, 1994) whilst working at General Motors, is 'the use or introduction of automatic equipment in a manufacturing or other process or facility'. There was of course automation long before 1936. Its seeds were sown with the development of metal machine tools in the Industrial Revolution. These facilitated the development of manufacturing, converting it from a craft-based task done by workers largely working by hand on a small scale, to one where products were produced in large factories. In 1873 the automatic lathe was invented, which allowed machines to be operated automatically rather than by semi-skilled workers. In the 1960s the author of this book worked at a small factory in the UK's West Midlands, setting up and programming automatic capstan lathes. One worker could easily look after four of these capstans, which would previously have required four semi-skilled workers. Automation effectively means, and implies, the replacement of human workers by automatic machines. The use of the term has now stepped out of the factory and we have, for example, office automation. Comparing those automated capstan lathes with industrial robots, one difference lies in the latter having an arm and a gripper that can be used on, for example, different cars as they pass through an assembly line. With capstan lathes there is no assembly line. But both, to differing degrees, replace humans.

2.3   THE MODERN AGE OF ROBOTS

We view the modern age of robots as beginning in the 1950s. Universities, particularly American universities, have played a key role either in the early advances in robotics or laying the ground for robots in other relevant scientific areas. As an example of the latter, the first electronic computer was developed at the University of Pennsylvania in 1946, weighing 30 tons. MIT quickly followed with a general purpose digital computer, which was the first to perform real-time computations and have a video display to show output. It was at this time that the Stanford Research Institute (SRI) was founded by a small group of business people. Some 20 years later, in 1966, the Artificial Intelligence Center (AIC) was set up at SRI and in the following six years they developed the unfortunately named Shakey, which was the first mobile robot to be able to navigate its way through its environment, with the aid of a television camera, a triangulating range finder and bump sensors. At about this time, a mechanical engineering student at Stanford, Victor Scheinman, developed a sophisticated robotic arm with a servo-controlled gripper with a working wrist and tactile sense contacts on the fingers. Scheinman left academia to further develop and then market his robotic arm, in the process founding his own firm. Other specialised research institutions followed. The Robotics Institute at Carnegie Mellon University, which was founded in 1979, developed the CMU Rover. Building on the principles of Shakey, it benefited from three pairs of omni-directional wheels, and had an improved camera that could pan and tilt.

But not all the work was done in universities. Two papers in 1950 by Claude Shannon, a mathematician from Bell Laboratories, on chess playing by machines have been linked with the birth of AI (Stone, 2005). It was General Motors who used the first industrial robot in 1961, Unimate, which unloaded high-temperature parts from a die-casting machine. General Motors then used 26 Unimate robots in car assembling. The Unimate robot was developed by Joseph Engelberger, together with George Devol, whom he met at a cocktail party, reflecting the importance of both chance and personal interaction in innovation. Devol had, in 1946, patented a device to control machines. In 1954 he applied for another patent. He claimed that his invention made available for the first time a general purpose machine that had universal application to a vast diversity of applications where cyclic control is desirable. He called his invention universal automation, or unimation for short. Together in 1962 they formed a firm, Unimation Inc., which by 1983 had developed into the world leader in robotics with 25 per cent of the global market share. As a consequence of his input, Engelberger has been called the 'Father of Robotics' (ibid.).

European firms too played their part in the development of robots. Faced with a shortage of workers, the Trallfa robot was developed in Bryne, Norway, to paint wheelbarrows. This was developed by Ole Molaug, a largely self-taught electronics engineer, and Sverre Bergene, a tool-maker. So successful was this that Trallfa decided to begin producing robots. Trallfa was bought by the ABB Group, a Swedish–Swiss multinational, and still operates in Bryne. In the UK, a British producer of agricultural machinery modified the Norwegian robot for use in arc welding. Another Scandinavian firm, Sweden's ASEA Group, now ABB, developed robots to do automated grinding operations and in 1975 there was the first installation of a robot in an iron foundry. To date the Japanese have played only a limited role in the narrative. But this began to change in the 1960s. Waseda University began the WABOT project in 1967 and five years later the first full-scale humanoid intelligent robot, WABOT-1 made its appearance. It could walk, grip and carry objects with its 'hands' and use its 'eyes and ears'. It also had a 'mouth' through which it communicated. This was followed in 1984 by WABOT-2, which could read music and play the organ.[7]

Much of this early development was focused on fixed robots to be used in industry. Non-industrial robots tended to be curiosities, possibly offering a glimpse into the future but with no practical current use. It is the industrial robots we have got used to seeing in the large car plants. They stay in one location and hence do not need to navigate their way through an environment. Early examples include both the Unimate robot and the Norwegian Trallfa robot. They tend to have a jointed arm and some form of gripper. Mobile robots, on the other hand, can move around their environment. Autonomous mobile robots (AMRs) can navigate their own way. Alternatively, automated guided vehicles (AGVs) rely on guidance devices that facilitate their travel in a relatively controlled space. These are a very different kind of robot to industrial ones and require different capabilities and present different problems for the scientists and the engineers to solve. Much of the development of robots as described so far sees the replacement of human labour when the work is difficult or unpleasant (as in an iron foundry) or impossible (as in space). This is so for many other uses. But increasingly robots, particularly with their increased mobility, are being used to perform tasks that, although they may be repetitive, are not dangerous or unpleasant as such, and in some cases the actions are not really repetitive.

Throughout this narrative the critical developments have come about through the drive and imagination of a few key people.

---

[7] https://www.youtube.com/watch?v=ZHMQuo_DsNU.

Sometimes they have been in firms and sometimes in academia. Universities and university academics have played a key role in the development of robots. The role of the university research centre has been of particular importance in this development, in focusing research on the specific area of robots and in bringing like-minded people with relevant skill sets together. Universities are also important in that robotics is the marriage of both AI and mechanics as well as several other disciplines. This is increasingly the case with modern technologies, which combine different disciplines together. But we have also seen the critical roles played by firms, sometimes very large firms, in developing robots. However, sometimes the firm has been the academic, or rather the academic has left academia to start their own firm. We have also seen how sometimes the research has been both specifically targeted, and at other times it was blue skies research that laid the basis for future developments. Increasingly firms are combining with university research centres, often part-funding the centre itself, sometimes funding a specific research project. To an extent this is changing the nature of both universities and firms as we shall discuss later.

2.4   ROBOTS IN FICTION AND MYTHOLOGY

In principle, robots have been around for thousands of years and abound in the mythologies of many countries. An early appearance of robots in history has been linked to Greek mythology, although it must be doubted whether the Greeks themselves perceived the concept of a robot as we know it. When Cadmus had founded Thebes, he killed the dragon that had been attacking his companions. He then planted the dragon's teeth in the ground, from which grew an army of armed men. Hebrew mythology introduces the Golem, a clay or stone statue, which is supposed to contain a scroll with religious or magic powers that animate it. The Golem performs simple, repetitive tasks, but is difficult to stop. Inuit legend in Greenland refers to the Tupilaq, or Tupilak, which is a creature created from natural objects such as animal bones, skin, muscles, etc. Ritualistic chants brought it to life, whence it was placed into the sea to hunt out and kill specific enemies. Thus, it is interesting to note that many of the early appearances of robots are linked with the military. This has often been the case, and remains the case today.

It is stretching reality a little to regard many of these as robots as we know them, that is, mechanically constructed machines made by humans, which can move and to an extent 'think'. These were beings brought to life by some form of magic or by the gods. However, by the end of the Middle Ages creations more recognisable as robots were beginning to appear in fiction. Edmund Spenser's *The Faerie Queene*, written in the late sixteenth century, featured an iron man, Talus, who worked alongside a knight, Sir Artegal, to defend equality and rights and punish lawbreakers, a task he performed without mercy or feelings. By the 1800s robots were becoming relatively common. *The Steam Man of the Prairies* by Edward S. Ellis (1868) featured a steam-driven robot to carry its maker around the American prairies. *Frank Reade and his Electric Man*, written by Luis Senarens in 1876, also featured steam-powered robot-like mechanisms. But perhaps the most famous of the pre-twentieth-century robot characters featured in Mary Shelley's (1818) *Frankenstein*. The robot Frankenstein created was a cut above its fictional predecessors in being able to learn for itself and, for example, to speak and read. This made it similar to AI and machine learning, but unlike AI, it was gifted with emotions. Many of these robots are somewhat clumsy and given to violence; a different genre is Alicia, a beautiful android, in Auguste Villiers de l'Isle-Adam's *L'Ève future*. Begun in 1878 and published in 1886, this story is often credited with popularising the term 'android'.

The term 'robot' first appeared in the Czech writer Karel Čapek's play *Rossum's Universal Robots*, first shown in 1921. The term was coined by his brother and stems from the Czech word robota, meaning 'serf labour'. However it was another writer who was to play a critical part in robot development. Isaac Asimov wrote a series of short stories on robots. These were published as a book in 1950 under the title *I, Robot*. A key element of these are his three laws of robots, although with the Zeroth Law, added in 1985, there are actually four laws. These are:

Zeroth Law: A robot may not injure humanity, or, through inaction, allow humanity to come to harm.
First Law: A robot may not injure a human being, or, through inaction, allow a human being to come to harm, unless this would violate a higher order law.
Second Law: A robot must obey the orders given to it by human

beings, except where such orders would conflict with a higher order law.

Third Law: A robot must protect its own existence, as long as such protection does not conflict with a higher order law.

These laws both reflect the hopes and the fears people have of robots. They also beg the question of who will police these laws? Who will ensure that robots behave in a way that benefits humankind, rather than damages it? The answer is of course no one, and in reality and for the foreseeable future it is not the robot that needs to be policed, but the humans who program them, and just as there are criminal individuals, and nations too, who will seek to damage others and break laws to their own advantage, so these individuals will make use of robots. But there is also the possibility that robots will cause damage, not out of malicious intent, but via Adam Smith's invisible hand. That is, the use of robots in manufacturing, services and in the home may have adverse impacts, not through the malicious intent of any one robot, but because their collective impact on society and the economy may be substantially or partially negative. That possibility is in many ways one of the key concerns of this book.

The advent of motion pictures brought to life many of these mythical and fictional creatures, as well as a seemingly endless supply of new artificial creatures. The 2015 film *Ex Machina* featured a humanoid robot in the form of a beautiful girl, Aviva, who entices and then betrays the person who falls in love with her as she escapes into the outside world. Robots have also featured in other films, beginning with Fritz Lang's 1926 film *Metropolis*. Maria is a humanoid robot in the likeness of a real person. She is so lifelike that perceptions that she is a robot only become apparent when she is burnt at the stake. In the intervening time, robot Maria has caused chaos throughout Metropolis, fostering murder and dissent amongst the downtrodden workers by her powers of persuasion, more than her actions. Maria is unusual in being one of the few 'female robots' in early science fiction.

In the 1951 film *The Day the Earth Stood Still*, Gort is a robot designed to protect citizens against all aggression by destroying any aggressors. He accompanies Klaatu to deliver an ultimatum to the people of Earth to pursue peace. Gort does not speak, but he can receive and follow verbal and non-verbal commands. He is an autonomous robot, who protects both himself, if it is right to refer to a robot as 'him', and Klaatu. Gort is also the pilot and captain of their spaceship. He is armed with a laserlike weapon that can vaporise tanks without harming their occupants. Since 1951 robots have featured in many films and television series and are often shown in a negative manner. The 1999 film *The Matrix* also takes an unfavourable view, not so much of robots but intelligent machines. They waged a war against the humans. When the machines' access to solar energy was blocked, they responded by using human bioelectricity as an alternative power source. The minds of the harvested humans perceived a simulated reality called 'the Matrix'. Their actual reality was much less pleasant.

In many of these films, robots and AI tend to be presented as a threat to humankind. They also tend to focus on robots in limited forms of activities – soldier, policeman, friend or lover. There are examples of autonomous vehicles (AVs), but seldom are robots depicted harvesting crops, caring for old people or cutting the grass. However, not all fictional robots present threats. In *Star Wars*, R2-D2 and C-3PO appear more favourably, not only being loyal to their human masters but to each other. C-3PO was designed as a protocol droid to assist with etiquette customs and, more usefully perhaps, translation. As often happens, fiction influences fact, and in 2012 the US navy built a robot called the Autonomous Shipboard Humanoid that was modelled on C-3PO. It is a fire-extinguishing robot and has similar movement abilities to C-3PO.

Fictional work is important in four respects. Firstly, it influences public attitudes. Thus it has been argued (Nevejans, 2016) that in the West, the influence of both early cultural references to robots such as the Golem, together with more recent fictional work, has caused people to view robots in a negative manner, which has hampered the development of the robotics industry. This fear is not so apparent in the Far East, where robots tend to be viewed more favourably. Secondly, fictional work can inspire the next generation of scientists, one example being Joseph Engelberger who, as we have already seen, has been called the 'Father of Robotics'. He started studying physics at Colombia University after being inspired by the work of Asimov. Thirdly, it provides inspiration and even templates for future developments. This is one area of science where fiction, both in books and films, has not only anticipated the future, it has helped shape that future. Fourthly, it provides a template to guide public policy. In highlighting, as it generally does, the worst of what can go wrong in the future, it highlights the need for corrective policies now, before robots have become a problem, rather than after the event.

2.5   ROBOTS TODAY

Roughly half of all the robots in the world are in Asia, 32 per cent in Europe, 16 per cent in North America, 1 per cent in Australasia and 1 per cent in Africa – 40 per cent of the world's robots are in Japan. However Japan does not have the highest density of robots, as can be seen from Figure 2.1. South Korea and Singapore have this distinction, followed by Japan and Germany. As an indication of market size, according to the International Federation of Robotics (IFR) sales of industrial robots in 2016 were 294 312 units along with 59 706 professional service robots. The latter saw an increase of 24 per cent over 2015. The worldwide market value for robot systems in 2016 was estimated to be 13.1 billion US\$, and 4.7 billion US\$ for industrial and service robots.

Figure 2.2 shows the growth of industrial robots over the period 1973–2020. This growth is rapid in recent years and accelerating.

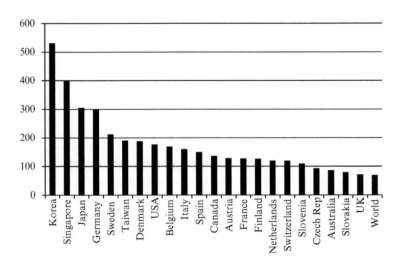

Source: International Federation of Robotics.

Figure 2.1   Robot density number per 10 000 employees in 2016

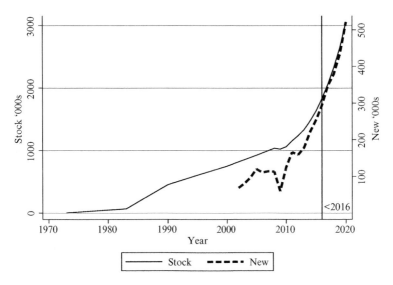

Source: International Federation of Robotics, data from 2017 onwards estimated.

Figure 2.2   Global industrial robots

The crisis of 2008 reduced new installations, but only briefly, and it appeared to delay rather than reduce new robots. Indeed, since the crisis, new robot installations have accelerated, which may be due to increased pressures to be competitive or technological developments, or both. Figure 2.3 illustrates how all the major areas of the world have been moving in tandem, apart from China, where, since 2013, the number of new industrial robots has accelerated substantially and is set to accelerate further in the coming years. However, China is a very populous country, of course, and in terms of robot density it does not rank that highly.

In 2016, the majority of new industrial robots were in just two industries: automotive industries (35 per cent) and electrical/electronics (31 per cent). But there are also a growing number of service robots and this might well be where the future growth potential lies. As Figure 2.4 shows, many are in logistic systems, that is, in both manufacturing and non-manufacturing environments such as warehouses. After that comes defence robots, robots in public relations (such as telepresence robots and robots for mobile guidance and information), and exoskeletons. Field robots are often found in

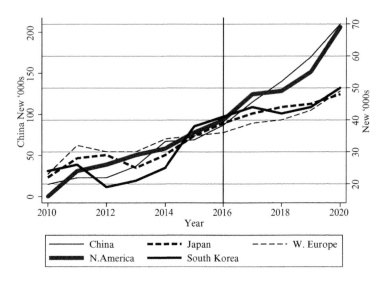

Source: International Federation of Robotics, data from 2017 onwards estimated.

Figure 2.3 New global industrial robots, selected countries

agriculture, for example, milking cows. There are also a large and growing number of robots used in the home for either entertainment or work purposes. Simply in terms of numbers, these household and entertainment robots dominate all others.

The IFR anticipates that during the period 2018–20 growth will be rapid. For example, 189 700 robots for logistic systems are anticipated to be installed along with 32.4 million household robots. The location of robot production is relatively widespread, although with different specialisations. According to the IFR, 54 per cent of professional service robots come from America and 27 per cent from Europe. Particularly important for the EU are field robots with about 91 per cent of the market in 2016, as well as 90 per cent of construction and demolition robots. America accounted for about 81 per cent of logistics robots and American firms dominate the household robot market. The Americas also accounted for the production of about 48 per cent of medical robots. But it was Asia/Australia, with 60 per cent of the market, who are strong in the entertainment market. These are, however, rapidly evolving markets and we can anticipate substantial changes in these market shares in the coming years.

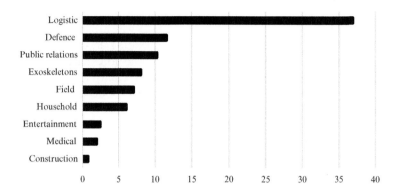

Source: International Federation of Robotics, figures in 000s units, apart from household and entertainment robots that are in millions.

Figure 2.4   Units of service robots (estimated for 2017)

2.6   CONCLUSIONS

Robots are as heterogeneous as is life on Earth. Although they are often compared to human beings, they are much more disparate than that. We have fixed (or industrial) robots and mobile robots, and the two are very different. Much of the early development of robots since 1945 was in industrial robots. Since then, anticipated in the work of da Vinci and in fiction, the focus has been more on mobile robots, and even more so on robots that, if not humanoids, at the very least resemble human beings. With mobile robots, there are further huge differences. As we shall see in later chapters, there are, or there are plans for, robots that try to resemble human beings as closely as possible. There are also snake-like robots that can be used to cross difficult terrain and can, being made of multiple parts, reassemble themselves and also easily repair themselves. There are robots that act in swarms, much like insects. There are very small robots that can be inserted into the human body to repair failing parts or repel damaging invaders such as cancer, and even smaller nanorobots. There are also robots that can operate under water and in space without an atmosphere.

It is difficult to think of any other technology that has been anticipated in fiction as much as robots. This was not the case for steam, electricity or the microchip. Stories about the horrors of the industrial revolution did emerge, but only when it was a fact, not when it was still to develop. Yet fiction has been leading developments in robots for over 100 years. The first android was foreseen in film in 1907. Over 100 years later it is still not a current reality, although it is clearly an evolving one. This alone suggests that robots and robotics are different to other technologies and that the robot revolution may be different to other technological revolutions. Fiction has viewed them as important enough for them frequently to figure in its work and that must be because robots have the potential, for good or ill, to fundamentally change our world and to change ourselves. Also, more than any other technology, fiction has influenced both the development of, and the public's attitudes to, that technology.

3. Robots now and in the future

3.1 INDUSTRIAL ROBOTS

There are two ways to review the different types of robots: firstly, to focus on the different types themselves and, secondly, to focus more on the different spheres within which they can be, or are likely to be, used. We largely follow the latter, although in doing so we inevitably discuss the former. We begin this review with industrial robots. They are used in factories and are typically used in industries characterised by large batch runs – indeed, runs that may last for years – without the need to reprogram the robots. Thus, in the automobile industry, they are dominant and have replaced large numbers of human workers. However, programming takes time and during this time the whole of the assembly plant can be at a standstill. When dealing with smaller batch processes, or when frequent small changes are needed to the product, this lost time is a problem. Robotics systems need to be as flexible as humans in coping with changed circumstances, if they are to replace humans in that part of manufacturing that is characterised by small batch runs.

Not all robots that work in an industrial setting are fixed. Increasingly there are some mobile ones. As in a warehouse, they can be used to move parts around the factory and offer the promise of another substantial step forward in manufacturing productivity, as well as the disappearance of still more manufacturing jobs. However, as with all mobile robots they need to be able to navigate their way around an environment, albeit a known environment. In a factory setting, mistakes or malfunctions can be costly and dangerous. But in saying this, it must be recalled that human beings also make mistakes.

3.2 WAREHOUSE ROBOTS

Compared to industrial robots, warehouse robots are relative newcomers. One prominent example of a company that has benefited from their use is Amazon. According to *The Seattle Times* at the end of 2016, the company had approximately 100 000 robots.[1] This has helped Amazon achieve a dominant position in the market and not just in the USA. Amazon's robots are built by Amazon Robotics. They are based on the robots of Kiva Systems, a company that Amazon bought for 775 million US$ in 2012.[2] These robots move product bins around the warehouses on a 'just in time' basis to humans who finish the packing process after which the boxes are shipped off to customers. When Amazon acquired Kiva they soon stopped supplying robots to other customers, giving Amazon a market advantage, although this is a gap in the market that other robotics companies are filling. Nonetheless this is an interesting aspect of how a company can affect competition by buying potential supplier companies, and raises competition issues that have largely not been picked up by the literature, nor by the regulators.

One of the main problems faced by warehouse robots is to navigate their way through the environment. This is often done through teach and repeat (T&R), that is, the manual creation and use of travel paths, thus removing the need for more complex simultaneous localisation and mapping (SLAM) (Gadd and Newman, 2015). T&R is often done by embedding infrastructure, such as barcode stickers in the floor, into the warehouse. The robot then uses these to navigate (D'Andrea, 2012). Alternatively, it is possible that this could be done by using cameras without any special infrastructure. During the teaching phase, the robot is taken along a route in part of the warehouse. The robot thus learns from human operators as they do their job. There then evolves a comprehensive and interconnected map of the warehouse developed over several such human-operator-led missions. During the repeat phase, the robot identifies its position, that is, it localises itself using a search process in the local graph neighbourhood. The robot can use this same search algorithm to autonomously navigate from one point to another, possibly over a path that it has worked out for itself.

---

[1]   http://www.seattletimes.com/business/amazon/amazons-army-of-robots-job-des troyers-or-dance-partners/.
[2]   http://www.zerohedge.com/news/2017-07-27/amazon-hosts-robotics-competition -figure-out-how-replace-230000-warehouse-workers.

3.3   AGRICULTURAL ROBOTS

Robots are deployed for myriad purposes in agriculture. For example, drones are used to monitor crops to ensure the efficient use of inputs like water and pesticides, and in forests to check the health of trees. More generally, robotics have been influential in agriculture in three main areas: (1) precision agriculture, that is, using sensors to precisely control the use of fertilisers and water; (2) the highly accurate use of tractors and combine harvesters in fields; and (3) the harvesting of fruits and vegetables. A major problem in the commercialisation of harvesting activities, however, is that robots need to be able to pick, and distinguish between, all kinds of vegetables. The variety of shapes, sizes and colours of just tomatoes, for instance, poses substantial challenges to a robot.

3.4   AUTONOMOUS VEHICLES

Driverless cars made an appearance at the 1939 New York World's Fair. In those days it was very much a concept, but now this concept is turning to reality. For example, Google's driverless car had been driven for more than 5 000 000 miles on city streets by February 2018.[3] Driverless cars are an example of an autonomous vehical (AV), that is, a vehicle that can be driven without human involvement. This is one extreme, but there are variations on this, and at a lesser level most of us are familiar with the automatic parking of cars. There are potentially many gains, including reduced road accidents, 90 per cent of which in the European Union (EU) are because of human error.[4] AVs may be linked to a central control system. The linking of several vehicles automatically moving together is known as 'platooning'. The central control system can provide information on future road conditions, with the platoon changing speed appropriately. This should improve traffic management in large urban areas and on major roads. It may also reduce fuel consumption and pollution (Mersky and Samaras, 2016). Thus safety and average speed are one gain. But obviously, too, travellers can spend their time in other

---

[3]   https://waymo.com/ontheroad/.
[4]   http://ec.europa.eu/transport/themes/its/road/index_en.htm.

ways than driving, in recreation, working or resting. Driverless cars also potentially allow people who cannot drive to travel (Fagnant and Kockelman, 2015).

But it is not a one-way street. There are various legal, financial, ethical, economic and technical issues that need further consideration and work before driverless cars become a reality for most people. Some of these are relatively simple technical problems, for example agreement needs to be reached on issues such as the technical standards for international compatibility and inter-operability. Other issues may be less easy to overcome. Kalra and Paddock (2016) emphasise that it is difficult to establish that AVs are safe before making them publicly available. AVs operate in a complex, dynamic and uncertain environment. Road signals are constantly changing, as are other road users. To compound these problems the vehicle may be moving at high speeds. Human beings are able to understand and interpret all of this, generally driving in a safe manner, whilst anticipating future developments. AVs will need to replicate these abilities, and more research is still needed for example in tracking and anticipating the movements of pedestrians, bikes and other vehicles (Katrakazas et al., 2015). There are other problems. Privacy may be threatened by the connectivity of the driverless car that potentially allows the continual tracking of movement. Increased road usage may increase, rather than reduce, the energy and carbon impacts (Wadud et al., 2016). But the most serious threat in terms of consequences is the possibility that criminals, terrorists or foreign governments may hack the system, causing, or threatening to cause, mass accidents (Grunwald, 2016). Meanwhile the development of AVs extends in reach. Thus, Boeing is looking into the possibility of pilotless commercial passenger jets, which fly using artificial intelligence (AI) to guide automated controls.

As with all other forms of robots we analyse, there is the potential for robots to replace jobs. With AVs the potential impact is particularly serious. With cars, taxi and delivery drivers are both at risk. When extended to larger vehicles such as trucks, the job replacement effects could be substantial. However, there are several differences in the impacts of driverless cars and trucks. Large trucks tend to avoid residential areas and they do not replace the citizen in driving their car, which many enjoy doing. But the damage an out-of-control truck can do is far greater than a car and there are many more jobs at risk with trucks than cars. The same is true with buses and coaches.

Finally, driverless trucks will probably reduce transportation costs and thus the cost of goods.

3.5 CARING ROBOTS

We now turn to examine some of the areas in which robots, generally called service robots, are being deployed or may come to be deployed in the future, beginning with caring for the elderly and other people in need of care. The elderly are an increasing proportion of the population in many countries. Providing care for them is difficult and costly and this is straining public finances. Robotics and related technology can help with these problems.

As is often the case in other contexts, the definition of a health-care robot is not entirely straightforward. At one extreme, general robots that assist older people with housework might be viewed as healthcare robots when they facilitate the older person living in their own home or help carers with cleaning duties in a care home. Robots can also provide companionship, and help monitor elderly behaviour and health. Robot care could potentially save money. They may also provide improved care. But there are potential disadvantages, problems and risks (Hudson et al., 2017a). Because the robot is in contact with the person much of the time, and because the robot is also in contact with some central control system, there is again a potential privacy problem. The robot can report on both conversations and actions. If they replace human contact this might enhance the social isolation of the elderly and the perception of a lack of control over their lives. These have repercussions, for example the former may increase dementia. At the current time there are also safety concerns. Robots are still less skilled than humans in avoiding obstacles, and collisions between humans and robots are possible. Other forms of injury include electrical ones. Compared to perfect human care, robotic care is almost bound to be an imperfect substitute. But human care is often not perfect, sometimes because of financial constraints and sometimes because of abuse. In this case people – even and perhaps especially elderly people – may find robots to be an attractive alternative.

Caring robots are developing rapidly and many of the problems with the current vintage will be solved with future generations. However, at this point in time, robots are still relatively rare in many contexts. PARO[5] is one commercially successful robot. It has the appearance of a seal and is designed to be a companion for an elderly person, who can hold and stroke it (Bedaf et al., 2015). In the future we can look forward to greater, and more beneficial, robotic involvement in caring for the elderly and generally in the home, including conversational robots. Apart from the direct benefits, in cutting costs, this can release money for other purposes, for example the quality of the accommodation, and the use of medicines.

3.6   MEDICAL ROBOTS

Robotic surgery frequently involves a surgeon operating remotely controlled robotic arms. In particular, this may facilitate minimally invasive surgery, that is, operations performed through small incisions. The surgeon controls the robot through a console. This can be from thousands of miles away and thus may eventually be of particular value in performing operations in relatively remote areas. The primary robotic system is the da Vinci system[6] and in September 2017 there were 4271 of these installed worldwide. Most of these were in the USA with a smaller number in Europe and Asia. They 'work hard' and in the USA over a half of radical prostatectomies and approximately a third of benign hysterectomies are done robotically (Tan et al., 2016). There is, however, relatively limited evidence that outcomes are currently being improved by robotic surgery (Porpiglia et al., 2013). Given that they are not cheap, their use is thus a little difficult to explain. One reason could be for marketing purposes, with potential patients impressed by the use of advanced technology (Boys et al., 2016).

Prices per unit vary from 1 million to 2.5 million US\$ (Barbash and Glied, 2010). There are also variable costs related to maintenance, including replacing some appliances following each operation. The training of surgeons may also take longer than training surgeons in conventional surgery. There is some evidence of medical benefits from robot-assisted surgery, although as noted earlier there is disagreement on this in the literature. The benefits include decreased blood loss and complication rates (Advincula et al., 2007). In part,

---

[5]   http://www.parorobots.com/.
[6]   http://www.davincisurgery.com/.

these benefits are due to the increased precision of robotic surgery, which might reduce the possibility of injuring blood vessels and other structures. On the negative side there is the possibility of robot malfunction. The evidence suggests that this is relatively rare and mostly takes place before the operation itself (Kaushik et al., 2010). Nonetheless, Alemzadeh et al. (2016) found evidence for 144 deaths, 1391 patient injuries and 8061 device malfunctions in the USA over the period 2000–13. There is also the possibility that teleoperated systems could be hacked to cause deliberate malfunction.

In the future, robotic systems will become more sophisticated and reliable, with surgeons more skilled in their use. These advances will expand what robots can do, and in doing so open up the possibility of new risks. Today, there are some examples of fully autonomous robots. However, these tend to be the exception, and typically the human surgeon is fully in charge of each step of the operation, with the robot the surgeon's junior assistant. This is very different to fully robotic surgery. Much of the research and development is looking into this or into robots that combine elements of both (Hoeckelmann et al., 2015).

3.7   ROBOTS IN EDUCATION

At this point in time robots are not widely used in education. This is partly due to expense, especially when many schools in many countries face substantial financial constraints, and also due to the need to educate teachers in their use and a lack of suitable educational material. Potentially they can be used as a learning tool, learning companion or peer, a learning instructor or teacher (Mubin et al. 2013). However, their role as teacher is minimal, as this requires better communication and autonomy than are currently available (Causo et al., 2016). Tiro is an example of a robot educational assistant. It has 16 pairs of ultrasonic sensors and 11 tactile sensors throughout its body. Its 'eyes' comprise two USB cameras, and it can be fitted with a laser range finder in its base for obstacle detection. Developed by Hanool Robotics Corp., together with four universities, Tiro has been used to check students' attendance and in providing conversation scripts, dancing, storytelling and role-playing. In its spare time, it has also been used to host wedding and other ceremonies.

Robots as tools can be found in all levels of education right up to university level. 'Edison' is one example. It is a cheap robot, which can detect light, follow lines and respond to sound commands. It is Lego compatible and can be expanded in size. Using Edison, students can learn about robotics and programming (Mondada et al., 2017). Apart from their use in computer science and programming, robots have been used in the teaching of music and languages (Causo et al., 2016). In order to act as an educational peer, it can be helpful if the robot has at least a semi-humanoid form, although often too they take the form of a 'pet'. For example, Nao and Asimo look like a small child (ibid.). In this role, robots assist learning by prompting or giving feedback as students go through a course, or by the student 'teaching' the robot. Children have been asked to teach Nao, in for example vocabulary, and in doing so, they themselves learn (Tanaka and Kimura, 2010). Nao has up to 25 degrees of freedom, which enables it to move, walk and even dance.

There are dangers in using robots in education, particularly companion robots and particularly with young or vulnerable children. Sharkey (2016) has discussed some of these dangers, emphasising the ethical concerns. There are the omnipresent concerns with privacy, particularly if the child comes to trust the robot, telling them things they would only tell a close friend. There are also concerns that robots could replace human contact, either because the child has a reduced need for such contact, or the robot is used as a substitute for humans in the classroom. Preferring to choose the robot's company over that of less predictable, and kind, human peers may reduce the learning of social skills. The child may always be able to tell the robot what to do, depending on its programming. This is not the case with other human beings. The child can also be nasty to the robot, possibly again depending on the programming, with no adverse consequence. This may affect their future behaviour. Sharkey suggests that the extent to which this is the case is simply not known. Finally, the child may become disillusioned when they realise that their robot friend is after all just a machine without feelings.

3.8   ROBOT SECURITY

There are already robots on the market to provide security in public areas such as shopping malls, airports, railways and schools. As an example, autonomous non-humanoid security robots patrol an area and report back on suspicious circumstances to a manually operated central control. Information is also deposited with the Cloud and can be stored for many years. Much of the technology is concerned with navigating the robot through populated spaces, with the robot having a sense of location. They come with sensors, thermal imaging, directional microphones and cameras that, together with 'anomaly detection software', helps them distinguish a potential criminal from an ordinary citizen. They can identify licence plates and smartphones including their MAC and IP addresses, and again all this information is transmitted to the Cloud. As well as providing continuous surveillance, they can also be used to patrol dangerous places and have been used in crime-ridden public car parks. They can also be used to detect non-criminal threats such as fires and gas leaks.

Thus, at the moment, security robots are aiding security officers rather than replacing them. They are the eyes and ears of security, rather than its physical enforcement. But clearly, as we move forward, the potential is there for them to take a more proactive role. The technology already exists for autonomous robots to kill or disable intruders. It is not so much the technology that is preventing that, but the ethics and the law. In future, as robots become more sophisticated, there is the potential for them to physically constrain people. Mistakes in doing this are one danger, as is, once more, hacking into the system. But there are also currently privacy concerns with data on individual movement continually being recorded. Even with surveillance robots there are dangers of them bumping into people and one recently fell into a fountain whilst patrolling an area. Small software or hardware flaws can have fatal consequences, as with incidents with autonomous helicopters, robotic cannons, and military robots (Lin et al., 2011). There has also been a reluctance to use fully autonomous robots to their maximum extent because the utilisation of AI with unconstrained learning algorithms can produce unpredictable reactions (Theodoridis and Hu, 2012). Because of this, the private sector has been hesitant to develop AI security robots. Nonetheless, this will change and Theodoridis and Hu (ibid.) describe recent advances in intelligent security robots as both illuminating and horrifying. Despite this, the potential benefits in fighting crime and ensuring public safety are considerable and it must be remembered that human law enforcement and security personnel also make mistakes.

3.9   ROBOT SOLDIERS

At present, the two most common types of robots used by the military are drones, that is, unmanned aerial vehicles (UAVs) and small unmanned ground vehicles (UGVs). UGVs are used for finding objects, such as explosive devices, and sometimes neutralising them. For example, the Dragon Runner[7] can be used to provide images of the terrain, either whilst mobile or stationary. It can also be used on sentry duty. UGVs, when armed with weapons, can also be used in a more offensive mode. Obvious political and ethical considerations are limiting their development in Western democracies, but less so in Russia where some have already been used in battle ground situations (Bogue, 2016). Thus, the MRK-27 BT is a radio-controlled mobile-tracked platform armed with a machine gun, grenade launchers and flamethrowers. It was developed at the Bauman Moscow State Technical University. These are more like AVs than humanoid robots. But the latter are also being developed. The Atlas humanoid robot is about 6 feet tall and was developed by Boston Dynamics supported by America's Defense Advanced Research Projects Agency (DARPA). It is able to drive a military vehicle, travel over rough terrain, climb ladders and perform various actions such as closing valves on a leaking pipe. It is claimed to be mainly for humanitarian uses, although again other countries may not be so reticent. Robot soldiers, including UGVs, might not only be faster, stronger and more reliable than human beings, they would also be immune from panic and sleep deprivation, and never be overcome with a desire for vengeance. On the other hand, robot soldiers might also be without human compassion.

UAVs are perhaps further developed in warfare than UGVs. They are also used for surveillance, but have been used to attack targets. Typically they are not yet completely autonomous. The General Atomics MQ-9 Reaper drone often has several people involved in a flight. However, in a new development, Perdix drones have been developed by MIT students.[8] They fly in cooperative swarms of 20 or more, communicating autonomously with each other and collectively decide on coordinated movements. They are given a task, and

---

[7]   https://www.qinetiq-na.com/products/unmanned-systems/dragon-runner/.
[8]   https://www.defense.gov/News/News-Releases/News-Release-View/Article/1044
811/department-of-defense-announces-successful-micro-drone-demonstration/.

decide themselves as to the best way to carry it out. Their potential is currently being explored by the US military. There are a number of potential uses, including surveillance and attacking enemy soldiers. Going even smaller, micro UAVs, which might be mistaken for an insect, could eventually be used in surveillance, in quite literally a 'fly on the wall' context. One variant of these sees real insects being transformed into cyborgs.

In the future, the types of robot that can be used by the military are almost as diverse as the number of different types of robots themselves. They include exoskeletons to enhance the abilities of human soldiers, self-driving submarines and fully autonomous helicopters. The ethical concerns, as well as more generally the dangers for human beings, are obvious and organisations such as Human Rights Watch have, for example, called on governments and the United Nations (UN) to outlaw the development of 'lethal autonomous weapons systems' (LAWS). In November 2017 the first meeting took place at the UN in Geneva of the Group of Governmental Experts on Lethal Autonomous Weapons Systems to examine these issues.

3.10   ROBOTS IN DANGEROUS SITUATIONS

Many of the skills and much of the technology used by military robots can also be used in dangerous situations. Teleoperated security robots are operated remotely by a human and are used in situations such as bomb disposal, fire extinguishing, remote surveillance and even in hostage situations. For example, after the earthquake that severely damaged the Fukushima Daiichi Nuclear Power Plants in Japan, teleoperated mobile robots were used to conduct investigations and measurement missions inside and outside the reactor buildings (Nagatani et al., 2013). Robots, including UAVs and robotic snakes such as those developed by Carnegie Mellon University, have been used in earthquake situations. At the moment they tend to be manually operated, but over time they will be more autonomous and able to make their own decisions. Thus, in the future, particularly in remote situations following a disaster such as an earthquake, hurricane or plane crash, rescue drones may well be first on the scene searching for, and offering medical care to, survivors.

3.11   MODULAR ROBOTS

Modular robotic technology is currently being applied in hybrid transportation, industrial automation, duct cleaning and handling. For example, military UGVs allow different superstructures to be mounted onto a vehicular platform, thus facilitating a range of different mission roles that can be performed. This development is still in its early stages. In some cases we see identical segments, which can be easily added to or removed. More generally, modules may have specialised functions such as that of a motor, gripping, sensory or energy storage. Some can adopt different configurations in order to perform various tasks in different complex environments. The components of modular robots also have the ability to perform tasks separately or jointly. The programming associated with modular robots, particularly with respect to the inverse kinematics and dynamics, is more complex than with traditional robots, as not only does the robot have to decide in which direction to move and what action to perform, but all the modular parts must contribute to these aims.

3.12   ROBOTS' CLOSE COUSINS

Bots have been referred to as robots without a body. Not all AI has a body, but bots and AI are similar to robots in that they are replacing humans, or at least the activities and tasks humans do, and indeed going beyond those activities. For example, AI is being used in a range of tasks that require pattern recognition in big data. This includes fraud detection (Phua et al., 2010), and is used by law firms in conducting pre-trial research with the scanning of thousands of documents, and also in health care with computerised diagnostic tasks. With large data bases, computers can diagnose and develop an optimal and personalised medical treatment plan by comparing each patient's symptoms, genetics, family and medication history, and so on.

The wearing of exoskeletons can enhance humans' physical abilities. For example, ankle exoskeletons can reduce the energy cost of walking by providing power at the appropriate time as the individual walks. Other exoskeletons allow workers to lift and move heavy objects with relative ease, and hence have potential use in a variety of workplace settings in factories, building sites, agriculture and mining. They can also facilitate movement in disabled or elderly people, and of course can be used to augment the power of military personnel. Rigid exoskeletons often include sensors in robotic joints that accurately track joint angles. But not all exoskeletons are rigid; for example, Harvard Biodesign Lab are designing sensors to measure suit–human interaction that can be integrated into wearable clothes. These and other wearable sensors can be used to help control the wearable robot or, alternatively, to monitor and record the movement of the wearer.

Brain computer interaction (BCI) enables users to instruct and communicate to a device purely through brain activity. They were originally designed to assist communication for physically challenged or locked-in users. But their potential use goes well beyond this. They tend to work in one of two ways. Firstly, via an electroencephalogram (EEG) cap worn on the head that detects neural activity using electrodes placed onto the scalp. Alternatively, wireless BCI involves a device planted directly into the brain. Neural signals can then be used to move a computer cursor and prosthetic limbs. With wireless BCI, work is proceeding on reducing the size of the computers to wearable dimensions. However, this work is still in its infancy and more research is needed until we arrive at the possibility of brain-controlled machines doing complex tasks such as tool use and better language communication. Nonetheless the DARPA in the USA has developed neural implants placed in sharks and cockroaches to remotely control their movement. Work is progressing on controlling other insects and also rats and pigeons. Such implants could technically change people into cyborgs. There is also the risk that such implants could be hacked and could then both be used to control an individual's actions and to extract information.

3.13 THE FUTURE

Robots will increasingly replace humans in their activities, but more than that they will increasingly do jobs humans just cannot do. In surgery for example, MIT, the University of Sheffield and the Tokyo Institute of Technology have built a prototype micro robot that performs simple procedures inside the stomach.[9] The patient swallows

---

[9] http://news.mit.edu/2016/ingestible-origami-robot-0512.

the robot, encased in ice, which makes its way to the stomach. The ice melts, and the robot moves under the surgeon's control. The robot can provide medication to an internal wound or repair it by covering it. The robot can also 'pick up' and remove foreign objects the person has swallowed. This has yet to be placed on the market, however in 2016 an academic at Oxford University used a robot to perform an operation inside a human eye.[10] Both these robots are controlled by a human surgeon, but it is possible that in the future surgical robots will be more autonomous. In the future too, and quite possibly the relatively near future, we can anticipate robots performing many more types of operations without the need for surgical incisions that are so damaging to human health.

One area where substantial growth can be expected is modular robots. As with much of robotics, the design problems are difficult and, with the building of many identical segments and joints, resource consuming. Despite this, there are some current applications and more in development. At the moment, there is relatively little autonomy. Tasks such as route planning, obstacle overcoming, stair climbing, and so on, can be performed autonomously. But in general, they tend to be teleoperated, although with some computer assistance helping the human operator. Self-configuring robots can autonomously change their internal structure for locomotion, manipulation or sensing purposes. Thus, a self-reconfiguring robot system could cross rough terrain using multiple legs and reconfigure into a snakelike shape to move through a narrow tunnel and on exiting this, transform into something else, suitable for the next environment.

Frey and Osborne (2017) document three problem areas facing robots in the immediate future. The first relates to perception and manipulation tasks, which they argue are unlikely to be resolved over the next two decades. Robots are still unable to match the depth and breadth of human perception, for example the identification of objects and their properties in an unstructured and complex environment. Robots also face challenges in planning the appropriate sequence of actions in order to move an object. Secondly, there are the creative intelligence tasks. At one level these require combining familiar ideas in a new, novel and useful way. This necessitates a large

---

[10]   https://www.technologyreview.com/s/603289/the-tiny-robots-revolutionizing-eye-surgery/.

knowledge base from which the robot, or bot, can learn. In principle, this is possible, and work is progressing to make it a reality. One robot artist, e-David, which was built at the University of Konstanz, uses an arm with a choice of five paintbrushes and 24 colours to create excellent paintings, including portraits. Robots are also capable of composing music, sometimes in the style of a known composer. Hence according to Frey and Osborne, the problem of generating creativity is not difficult, rather the main problem lies in formulating creative values with sufficient clarity to be programmable. The third and final problematic area relates to human social intelligence. Some aspects of human social interaction can be programmed, but the real-time recognition of human emotions is difficult, as is the appropriate response to such emotions – although of course the latter can be the case with humans themselves. Frey and Osborne argue that it will take a decade or two to make serious progress with these problems, but when we do, the employment impact is likely to be vast. Some might argue that two decades is a long way ahead, but to my way of thinking it is not that long a time to prepare for such an impact.

APPENDIX 3.1   SOME DEFINITIONS

We begin with a taxonomy of robots. *Robotics* deals with the design, construction, operation and application of robots. A *robot* is a machine that can carry out actions automatically. *Autonomous robots* are intelligent machines that can do these actions without explicit human control. *Mobile robots* are not fixed to one physical location – this is the type of robot that typically populates science fiction films. There are many types of robots, and in some respects it is more appropriate to compare them with life on planet Earth rather than just humans, albeit with the qualification that snake robots, for example, may have the thinking ability of humans rather than snakes.

Industrial robots   These are used in industry, particularly, for example, on automobile production lines. At this point in time most industrial robots are fixed with powered mechanical arms with the ability to perform humanlike actions. They generally have a jointed arm and an '*end effector*', that is, a device at the end of an arm to which tools or a gripper may be attached. These grippers can be of several forms including claws, which physically grasp the object the robot is working on, and pins, which penetrate the object. They are usually fixed, although this is not always the case, as the ISO 8373 standard makes clear in defining a manipulating industrial robot as 'an automatically controlled, reprogrammable, multipurpose, manipulator programmable in three or more axes, which may be either fixed in place or mobile for use in industrial automation applications'.

Service robots   ISO defines these as a robot 'that performs useful tasks for humans or equipment excluding industrial automation applications' (ISO 8373).

Mobile robots   as their name implies can move around. They can be autonomous or rely on guidance devices that facilitate travel along a pre-specified route in a fairly known space. Mobile robots are, for example, used in warehouses to take objects from shelves to be packed. In reality, almost any robot that is not an industrial one can be classified as a mobile one, and thus they encompass a huge range of activities. The *automated guided vehicle (AGV)* is a mobile robot that navigates its way through an environment by following markers or wires on a floor, or uses vision or lasers.

Collaborative robots (cobots)   are robots that interact with humans whilst performing tasks. Because they work with humans, often in an industrial setting, safety is a prime concern. Nonetheless it is an area that both the EU and the American government are encouraging in both research and development contexts.

Modular robots   are a system consisting of many modules that can alter their construction to perform different tasks in different environments. In effect a modular robot is a combination of smaller robots into a larger one. They are characterised by a large degree of adaptability and an ability to self-repair. Their development is still in its infancy. There are difficulties in programming them as not only do you have to program each module, but you also have to program the communications between each part. There are complex programming problems but potentially they are very flexible, easy to repair and to reconfigure to perform different tasks. If a human breaks an arm, then it will be potentially several months before they work again and possibly never as effectively as before the break. With a modular robot it is a simple matter of replacing the broken component with another identical one.

Humanoid robots are built to resemble the human body. In general, they mimic the human exactly, with a torso, a head, two arms and two legs. Some even have human facial features with 'eyes' and 'mouths'. The human design is in part for functional purposes as the human body is an efficient structure for many activities.

Androids are humanoid robots built to aesthetically resemble humans. An example is Erica, developed in Japan by Professor Hiroshi Ishiguro from Osaka University's Intelligent Robotics Laboratory.[11]

Nanorobotics is a very young technology, creating robots made up of components on a nano scale. At the moment, they range in size from 0.1 to 10 micrometres. One example of their use is as a sensor, able to count specific molecules in a chemical sample. Other potential uses are in nanomedicine, for example destroying malignant cells in the human body, and detecting toxic chemicals in the environment. But their uses may well extend much further in the future. Again much of the work is being done in universities. Hence the University of Manchester has created a nanorobot to complete basic tasks in the lab including the building of molecules. To give an idea of scale, a billion billion of these robots would still only be the size of a grain of salt.

Field robots work in unstructured, and frequently unknown, environments, such as in the air or under water, in mines, forests and on farms.

Telerobotics relates to the control of semi-autonomous robots from a distance. This is often done using wireless connection. It combines teleoperation and telepresence. *Telepresence* allows a person to have the appearance of being present. *Teleoperation* involves the distant

---

[11] https://www.theguardian.com/technology/2015/dec/31/erica-the-most-beautiful -and-intelligent-android-ever-leads-japans-robot-revolution; https://www.youtube. com/watch?v=VD2Btrj9ipY&index=1&list=PLG7sRAdtlqAlkwFmOR26occCJzbF NrymM.

control of a machine, that is, a form of remote control. It is not just restricted to robotics.

Some robots have specific names, for example unmanned aerial vehicles – better known as *drones*. Apart from military uses, drones are also used for surveying in construction, inspections for real-estate sales and evaluating infrastructure. News-broadcasting organisations are also beginning to use them to film events. At the moment, drones cannot react to others or interact with their surroundings. Some current drones can fly on their own, but they fall a long way short of what is required for safety. Future drones will sense, think and act to avoid collisions and safely achieve their intended purposes.

Kinematics is a branch of physics and classical mechanics focused on the spatial position of bodies, in this case the robot, the rate at which it is moving and the rate its velocity is changing. It ignores the causal forces of this motion. *Inverse kinematics*, with respect to robots, deduces the robot's required movements to fulfil a specific task from sensor-obtained data, giving for example position. Use is made of the kinematics equations to provide a desired position for each of the robot's joints, for example 'the elbow'. Inverse kinematics thus transforms a motion plan for the robot into trajectories, that is, joint positions, velocities and accelerations. Once these have been calculated, the next step calculates the forces and torque required by the joint actuators to move the links. This process is known as *inverse dynamics* (Marothiya and Saha, 2005).

Robots have many close relatives that will impact on people in, to some degree, similar ways to robots. One close cousin is a *bot*. A bot is an autonomous program on a network that can interact with systems or users. They are usually powered by AI, hence the name, as in 'robot'. But they may also rely on humans. In many respects a bot is a robot without a 'body'.

Exoskeletons in robotic terms are wearable devices that augment human performance. They can be powered and equipped with sensors or they can be entirely passive. They can cover just a part of the body, for example an ankle or the upper part of the body, or cover the whole of the body. Unlike robots, exoskeletons do not replace people, rather they use robotics and biomechatronics to enhance human abilities. *Biomechatronics* is the combination of biology, mechanics, electronics and control.

Cyborgs The word cyborg was first coined by Clynes and Kline (1960) as an abbreviation for 'cybernetic organism', although the concept can be found in the literature long before that. It is a being, possibly a human, with both organic and biomechatronic body parts. A cyborg is basically a combination of a being and machine in which the control mechanisms of the being part are modified so that the being can live in an environment different from the normal one. An example is where humans are adapted to live in environments such as space where there is less oxygen and more radiation than on the Earth. However, many applications of cyborgs simply refer to the being, for example an animal or insect, being controlled from the outside by humans.

4. The science of robots

4.1 INTRODUCTION

One cannot really understand what robots are capable of, either now or in the future, unless one has some understanding of the underlying science. There are three main elements to robots: (1) manoeuvrability; (2) a sense of space or location; and (3) an intelligence to guide the robot's actions. Manoeuvrability includes both the ability to physically change location and to move arms and use them to grasp objects. In doing this, robots build upon the fields of mechanics, automation, electronics, solid-state physics, fibre optics, computer science, cybernetics and artificial intelligence (AI). In addition, robots used in specialised activities, particularly service robots, need contributions from the relevant disciplines, for example mathematics, logic, psychology, the law, biology and industrial design. The wide range of expertise that is being brought into play in the development of robots in part differentiates it from previous technological revolutions.

In order to understand this, think of the tasks confronting a floor-cleaning robot. It has to move around a space, which may or may not be known to it, identifying dirt, cleaning the area that needs to be cleaned, but without damaging it. In order to do this, it will need variable suction power, for example increasing the suction with particularly stubborn dirt. It also needs to cover the whole of the cleaning space and avoid obstacles when moving. In this case, it is not necessary that it identifies the obstacles as chairs, desks or cabinets; it simply needs to avoid them. This involves both movement and obstacle sensing. It also needs an understanding of what the space is, so that it can be fully covered, and whether or not it has cleaned that space already. A companion robot, on the other hand, eventually needs to be able to navigate in the same manner as a cleaning robot, but also be able to identify a human, and not just that the person is human, but be able to put a name to them. It also needs to be able to converse with the human, that is, to 'understand' what the human being is saying and to reply in an appropriate manner. A driverless car may also need to converse with its passengers, but its main task is to identify where it is in an uncontrolled and dynamic environment, drive to where it has to go to, and in the process avoid crashing into other vehicles, people and obstacles. Here, then, we have the basic tasks of a robot: movement, location identification and an intelligence – although the intelligence varies from fully autonomous robots to human-controlled and pre-programmed ones.

4.2 MANOEUVRABILITY

This is the simplest of the three basic tasks and the one on which people have been working longest, although there are still problems to be overcome. There are two types of movement. Firstly, of a robot arm or wrist whilst the robot is stationary and, secondly, movement of the actual robot itself. In industrial robots, manoeuvrability is particularly important with respect to the robot arm. The degrees of freedom (DOF) relate to the number of single-axis rotational joints. The more joints there are, the more flexible is the robot. There are six DOF if the robot has six movable joints. This is the minimum to enable the robot to move in all directions. Two DOF permits a robot to move an object anywhere in two dimensions and put it down again, but without changing the object's orientation. Three DOF permits similar movement in three-dimensional space, whilst four DOF allows the robot to change the object's orientation along one dimension. *Six-axis robots, that is with six DOF, are the most common type of industrial robot.* They can place an object at any location within their reach, and can orient the object in any direction. They can similarly work with a tool if this is placed at the end of the arm. If the industrial robot itself can move, for example along a rail, then it is referred to as having seven or more axes.

Mobile robots can move bodily through space, not simply along a rail. They are to fixed robots what animals are to plants. This movement can be in two or three dimensions and on wheels, legs or tracks in as diverse a manner as that of life itself. Snake robots are made up of segments, each may move on wheels, tracks or legs. They are designed to be able to move across difficult and varied environments such as construction sites, across gaps and inside pipes. For example, they can cross gaps by stiffening the joint servo-mechanisms and can also cross wet ground by softening them (Granosik, 2014). Mobile robots often have more DOF that can relate to the legs, the neck and the torso as well as to each arm.

4.3   NAVIGATION AND AWARENESS OF POSITION AND LOCATION

Industrial robots are typically fixed, and for them the problems of location identification and navigation are much smaller than for service robots, which are typically mobile and frequently operating in an unknown and potentially difficult environment. If movement is undertaken autonomously, the minimum task is to reach a certain point whilst avoiding obstacles. For social robots this may need to take into account human, or possibly animal, sensitivities that may differ from culture to culture, for example the appropriate distance to pass by a human or animal without invading their personal space. Thus, passing within three inches of a human being is often not acceptable, but nor is avoiding them by 50 metres. The ideal is that the robot will follow a similar route to a normal human being of that culture. This process of navigation and location identification revolves around two types of technology. Firstly, sensors provide information on the immediate environment; secondly, AI interprets that information and uses it to build maps and construct plans of future action, for example a projected route to reach a destination.

Early work on autonomous navigation focused on avoiding obstacles in a relatively random manner. A metric map is a step forward from this and permits, for a previously unknown space, the production of a 3D map. Simultaneous localisation and mapping (SLAM) is the process of making or updating a map of an unknown area (the mapping part) whilst keeping track of the robot's location within this area (the location part). It is used when the robot is operating in unknown spaces, for example with autonomous vehicles (AVs) and robots implanted into the human body. The SLAM algorithm focuses on three main activities, involving: (1) sensors; (2) data processing; and (3) map representation. Sensors, such as cameras, microphones and sonars, obtain data, which can then be analysed using algorithms based on techniques such as filtering, smoothing and AI. Other sensors can be used, for example tactile sensors derive information on temperature from physically interacting with the environment.

There are many different types of maps that robots may construct. Grid or location maps, either in two or three dimensions, use a matrix of cells to represent the environment. Occupancy grid maps (Elfes, 1990) represent each small rectangular section of the environment with a cell, which is modelled with a binary variable indicating the probability of an object in that section. The map representation builds a model of the environment together with the position and route of the robot. Topological maps represent the environment through a set of compact and connected paths and intersections (Kuipers and Byun, 1991). An example of such a map is when pigeons use highways and highway intersections to navigate over long distances (Guilford et al., 2004). These are not sequential processes, with data processing and map representation interacting. Semantic maps contain information on the functions and relationships of the objects, for example the humans, walls, chairs and doors forming the environment (Nüchter and Hertzberg, 2008). They are similar to topological maps, but convey more detailed information on the environment. The robot is not restricted to any one type of mapping method, and may combine several together to form a hybrid map.

When multiple robots act together, we need to move to multiple-robot SLAM, which is far more complex. Not only must each robot know where it is and where it is going, it must also know where the other robots in the team are, where they are going and how the sum total of each robot's movement contributes to the achievement of the objective. Against this extra complexity, multiple-robot SLAM allows tasks to be completed more efficiently and is also robust against the failure of any one robot.

4.4   GUIDING INTELLIGENCE

The intelligence of a robot is in essence a software system consisting of a set of modules – that is, programs – performing specific tasks. One such task may process human speech and another recognise objects in the images captured by its video sensors. These relate to the robot understanding its environment. Another aspect of intelligence,

which is common to all robots to differing degrees, involves forming
a plan of action. The robot has a target or a goal, and must have a
plan to achieve that goal. The goal may be to efficiently clean an area.
The robot must know both what its next movement will be and how
that contributes to achieving the overall goal. When there are other
agents who can influence the outcome, the robot must be able to
make decisions under a potentially new and changed situation. Thus,
they must be able to assess the environment, make predictions, evalu-
ate the consequences of those predictions and behave accordingly.
With cleaning or moving an object from one location to another, the
goal is reasonably clear. For an industrial robot the task is simpler
still. But this is not always the case. The goal of a caring robot is to
look after and provide company to a human. The robot will need to
take actions that facilitate the achievement of this overall goal. In this
case, a 99 per cent success rate is not good enough, as the 1 per cent
failure rate can have significant adverse consequences.

Different robots have different tasks and abilities. Caring robots
need to not only understand speech and the meaning of words, but
also the context in which they are delivered, and that is difficult.
People can say the same thing, but mean something totally different.
'I really want to go to that cricket match' may be said sarcastically
or truthfully. People can also say things in a moment that they do
not mean. 'Will no one rid me of this turbulent priest?' may or may
not have been said in frustration with no real intent by Henry II of
England, and again robots would need to detect that, even if his
knights could not. For this and other reasons, social robots will need
to be able to detect mood and emotion. This will, amongst other
things, require an empathy module to analyse facial cues, and the way
the speech is delivered. Affective computing relates to recognising,
understanding and simulating human feelings and emotions. Apart
from facilitating interaction with humans, this is important in being
able to predict a human's actions by understanding their emotional
state.

4.5   ARTIFICIAL INTELLIGENCE

Not all robots have AI, some are simply pre-programmed or, for
example, teleoperated. AI relates to a machine mimicking the intelligence of human beings with respect to activities such as 'learning'

and 'problem solving'. This involves acquiring a knowledge base, the ability 'to reason', make deductions, take actions and possibly to plan ahead. AI began as an academic discipline in 1956 and has gone through periods when it went out of fashion. Because it tries to mimic and to an extent replicate human thinking, psychology has played a part in its development, as of course has computer science. Other disciplines have also contributed, including statistics, mathematics, neuroscience and economics. Humans tend to use heuristics to make quick intuitive judgments, rather than engage in detailed reasoning. Probability theory can be used to replicate human guesses, although with humans such guesses are often based on non-probabilistic intuition. It is, however, debatable whether we want robots to replicate that.

As with humans, machines cannot make decisions in a vacuum: they need a knowledge of objects and relations between objects, situations, events and time. If their sensors provide information on an object they need to be able to identify it and that requires a knowledge of objects in order to be able to categorise it. There are many problems relating to this. People tend to use working assumptions. If people talk about a bird, they have in mind a smallish creature that flies. But this needs to be qualified as not all birds are small or fly. This is known as the qualification problem (McCarthy, 1977). It creates problems for AI in devising rules or working assumptions applicable in all contexts.

This knowledge is acquired through machine learning (ML), which can be supervised or unsupervised. The former is when the outcome is known and the task is to get the robot to achieve that outcome. The robot is provided with the knowledge. Unsupervised ML is more complex, as it is applied when the outcome is not known. This relates to the ability to solve problems when the answer is not known, simply using the input data. The difference is similar to the student being taught things by a teacher and a researcher discovering new knowledge. It is also the difference between a knowledge of a known space or environment and that of an unknown space. Unsupervised ML relates to finding patterns in data. For example, the problem is to categorise geometric shapes by their colour and shape. With supervised learning we teach the machine that a triangle has three sides and a circle has no sides. It can also be taught to recognise colours and that a colour with certain characteristics is called, in English, blue. In unsupervised ML the machine is given the task of categorising the shapes and then it will seek patterns. It will recognise that circles have similar shapes and classify them together and assign a label of its own to them. If it needs to communicate with other machines they may develop a common language, and one that humans may not understand. Having classified the shapes in this manner, sub-categories may be created on the basis of colour – although the machine may decide to categorise on size instead. Mistakes are made, just as humans make mistakes, but sometimes, as with humans, the machine can learn from these mistakes, to make better decisions in the future.

Apart from classifying objects, ML also uses regression techniques to estimate the functional relationship between inputs and outputs, thus allowing it to predict how a change in inputs should impact on outputs. This can be used when pricing options in finance. When the response is not as expected, this is information the machine uses to improve its performance. If, for example, the machine is trying to translate German into Spanish, it will begin in a random manner, and compare its first somewhat poor attempts with actual translations. The program learns what worked and repeats the process. After each iteration it improves and eventually after an enormous number of attempts – which may not take it very long – it reaches and may even exceed the level of human translators. A validation data set is then used to compare the accuracy of different algorithms. Finally, using the optimal algorithm, the robot's abilities are then tested with a set of test data. The test data are independent of the training data, but with the same statistical properties.

Brynjolfsson et al. (2018) argue that ML is a general purpose technology (GPT) and hence has substantial implications across industries. Recently progress has been driven by techniques involving deep learning, or more specifically a class of algorithms called deep neural networks. In part this progress has been facilitated by these new algorithms, but also by enhanced computer power and speed, together with huge data storage abilities and a massive amount of data. The storage facilities are often linked to the Cloud and the data may be collected from social media, Internet search engines, e-commerce platforms and so on. This has allowed machines to work, and work very well, in tasks involving predictive analytics, and image and speech recognition. Past automation used explicit computer algorithms, written by a programmer, and thus were limited to activities that could have been codified. They thus suffered from

Polanyi's Paradox (Polanyi, 1966), that is, that our tacit knowledge of how the world works often exceeds our explicit understanding, and hence we cannot codify all we know. ML overcomes this by, in effect, the machine developing its own tacit knowledge. This substantially expands the set of possibilities that can be automated.

Artificial neural networks (ANNs) are vaguely based on the neural networks in the human brain, with neuron nodes connected together. But, unlike the brain, where any neuron can connect to any other neuron within a neighbourhood, ANNs have discrete layers, connections and directions of data movement. Take the example of fraud detection: the first layer of an ANN processes data input such as the amount of the transaction, the geographical location, the spending patterns of the individual and the product being bought. These data are then all passed on to each node in the second layer. Each piece of data is analysed separately in how it bears on the problem of deciding whether the transaction is fraudulent or not. This second layer has a number of neurons each combining the information in different ways, from a different perspective, and potentially passes the outcome on to a third layer and so on. This continues across all levels of the neural network. The final output is then based on all intervening outputs. Because they are analysed separately, ANNs can readily handle non-linearities. The layers between the input and output layers are known as hidden layers. A deep network has several, possibly many, hidden layers, which is why it is termed deep. There is considerable work going on as to how these different layers combine. The calculations involved can be very large and need to be done in parallel, rather than sequentially, which is how the typical central processing unit (CPU) works. In analysing an image, for example, there may be millions of pixels to interpret in deciding whether the object is a human. To begin with, each node makes a decision on its bit of the data, which in subsequent layers may be combined with adjacent ones. CPUs with multiple cores help, but are expensive. Graphics processing units (GPUs) have a processor with thousands of cores capable of performing mathematical calculations in parallel, and it is these that are used in deep learning applications, and that are providing the enhanced computer power referred to earlier.

Frey and Osborne (2017) claim that ML avoids the bias that may characterise human decision-making. Others are less sure. For example, Kulshrestha et al. (2017) argue that bias in Web search engines could influence public opinion by preferentially ranking results that correspond to one particular perspective above others. Such bias could be deliberate as the search engine ranks a firm's own products more highly than competitors', or it could be accidental. Bias can also impact on the outcome of elections and votes. But bias can also arise from AI finding patterns in data that represent what we understand as bias. This may be particularly likely if the training data is based on actual human decisions, which in themselves are biased. AI can be used to make or guide decisions on, for example, loan applications, and bias there can lead to one part of the community being discriminated against. This is also the case in using AI to decide who to select for a job interview and whether to grant a prisoner parole. The problem is, we know how and why we as humans make individual decisions, but we may not be certain as to why a machine may be making decisions, that is, what patterns it found that informed the decision. Thus to unquestionably take the machine's advice poses a risk.

4.6   INTEGRATING THE THREE STRANDS

The three aspects of the robot work together in the 'perception to action' process. The sensors provide information to the processing unit of the robot, which it then uses to understand its environment and its location in that environment. Action follows after the robot has evaluated that information. This requires some form of decision-making process, a process that may be based on ML, probability theory and reasoning, or some combination of these. But just as it is the human brain that decides on the action to perform and it is the parts of the body that carry out the action, so it is the thinking part of the robot that decides on the action, and the body of the robot that implements that action, and this is where mechanical engineering becomes important. Then, immediately following the action, new information may be received by the sensors, which is again transmitted to, and evaluated by, the processing unit. The consequence is that the action may continue as planned, or some form of correction may be made. These areas are developing all the time. Technological innovations, such as the miniaturisation of computers and, on the software side, mathematical control theory, together with improved sensor technologies, have substantially improved the ability of robots to autonomously achieve their targets.

4.7 PHILOSOPHICAL QUESTIONS

We finish this chapter with some reflection on certain philosophical issues that have surrounded the development of robots. Firstly, can robots think? This was the question posed by Alan Turing at the beginning of his 1950 paper (Turing, 1950). It was a question he chose not to answer, instead turning to the different question of whether machines can do what humans as thinking entities can do. The Turing test, developed in the paper, proposed that a human evaluator would judge a text-based conversation between a human and a machine. If the evaluator could not distinguish the machine from the human, the machine passes the test. His paper also considered nine opposing views to his own, which encompass all the major arguments against AI that have since been discussed in the literature. The test has been influential, even though subject to much criticism. People have questioned whether relying on the evaluator's judgment is a suitable basis to compare a machine with a human. Thus Searle (1980) argued that the machine could pass the Turing test simply by manipulating data, such as symbols but with no understanding, and without this there can be no 'thinking' as we understand the term. Hence the Turing test cannot be used to determine whether a machine can think. This type of consideration has led to a debate on the nature of intelligence. In passing, we note that in 2018 the technology known as Google Duplex was able to hold a telephone conversation to book a restaurant table without the person at the other end of the conversation realising it was a robot, which some have interpreted as passing the Turing test, although by being voice-based rather than text-based it arguably went further than the Turing test. Related to this is the concept of self-awareness. Bringsjord et al. (2015) argue that it is impossible for a machine to have true self-consciousness. But nonetheless they present evidence that some robots have shown a degree of self-awareness, and describe a case where a robot became aware of itself, and corrected its answer to a question once it had realised this.

A further consequence of this is that the literature has begun to consider whether robots deserve legal rights (Coeckelbergh, 2010). In evaluating this discussion it is important to bear in mind that it relates to robots as they are now and in the foreseeable future. At this point in time it must be doubted whether robots will develop a degree of consciousness or self-awareness to justify legal rights in the medium term, although there may be other reasons to give them such rights. However, it is difficult to see that far into the future and today's conclusions may be less valid in 30 years' time as robot abilities expand. Thus, even if legal rights for robots seem irrelevant today, this might not always be the case for at least some robots.

5. The impact on employment, unemployment and wages

'We are being afflicted with a new disease of which some readers may not have heard the name, but of which they will hear a great deal in the years to come – namely, technological unemployment.'
(John Maynard Keynes, 1933)

'Any worker who now performs his task by following specific instructions can, in principle, be replaced by a machine. This means that the role of humans as the most important factor of production is bound to diminish – in the same way that the role of horses in agricultural production was first diminished and then eliminated by the introduction of tractors.'
(Wassily Leontief, 1983)

5.1 INTRODUCTION

The statement by Leontief, exceptionally perceptive given when it was made, is not one widely shared by many economists, at least until recently. The common view amongst economists, for example Keynes (1933), has tended to be that although there may be short-term threats, in the long run these will not prove to be the case. But increasingly, in recent years, with the growth of robotisation and the growing sophistication and abilities of robots, some economists are beginning to wonder if finally the fears of Ned Ludd and the Luddites are about to come true. One catalyst for such fears is the recent stagnation of wages, particularly in the USA, and a growing inequality within developed countries (Benzell et al., 2015). Thus, in the USA, median wages have changed little since 1970. In the UK there has been little wage growth for the average worker since the beginning of the new millennium. Apart from robotisation and automation, this has also been blamed on immigration and globalisation.

These fears centre less on unemployment than inequality. This is a consequence of the perceived uneven impact of robots on the labour market, with the low- and medium-skilled workers losing and the higher-skilled and better-educated gaining. But is this necessarily the case with robots becoming increasingly able to replicate more skilled work? This is one of the issues we will explore in this chapter. We focus on two issues in particular. Firstly, the possibility that robots are causing unemployment and, secondly, that they are adversely affecting the living standards of at least some people in society, possibly widening inequality. We shall approach this by first looking at individual perceptions of the impact of robots and their abilities. Secondly, we will focus on actual unemployment and prosperity. In doing this we build on the work by Hudson et al. (2017b).

5.2   THE LITERATURE

There is concern amongst economists about the impacts of robots and artificial intelligence (AI) on the labour market. However, on balance the view amongst economists is that they will not cause unemployment, although they may lead to greater inequality and some people may experience declining living standards. Some economists are even of the view that, as with previous technological revolutions, on balance in the long run this one will be favourable in terms of its impact on the economy and individual workers. But it is something we are struggling to analyse. On the empirical side there is a lack of data, particularly time series data, and on the theoretical side economics has not developed a satisfactory framework to analyse the impact of automation on the labour market (Acemoglu and Restrepo, 2018).

The robot revolution tends to be capital intensive, labour saving and skilled biased (Roubini, 2014). It will favour the better educated and more skilled workers, but reduce aggregate employment in the economy. As argued by Janoski et al. (2014), automation and robotics reduce jobs on assembly lines, whilst creating a few more jobs in robot design and maintenance. Peláez and Kyriakou (2008) concluded that automation significantly transforms organisations, resulting in companies with a high level of robotisation tending to hire fewer, but more skilled, employees. On the other hand, Autor (2015) argues that automation, as well as substituting for labour, also complements it and increases output, thus paving the way for a higher demand for labour, neutralising any adverse impact on unemployment. However, he also emphasises that, although aggregate wealth will increase, the fairness of its distribution could be harmed.

The empirical work does tend to find some negative impact on unemployment, but this is sometimes restricted to manufacturing and compensated for elsewhere in the economy. It also often finds an adverse impact on the wages of some workers, but less so the highly skilled. Thus, Autor with others (Autor et al., 2003; Acemoglu and Autor, 2011; Autor and Dorn, 2013) link recent declines in the employment and wages of middle-skilled workers to the development of AI. However, at the beginning of a major technological revolution we often observe such disruption to the labour market. For example, Katz and Margo (2013) show a similar polarisation of labour in the early stages of America's industrial revolution, as do Goos et al. (2010) for Europe. Thus deciding whether these declines will be permanent or just temporary is difficult. From a slightly different perspective, Schmitt et al. (2013) conclude that the timing of changes in relative wages and employment exonerates robots as the cause of post-1970s US job polarisation.

Others are less certain and Acemoglu and Restrepo (2017) find that, for the USA, in the areas most exposed to robots, the introduction of a new robot per 1000 workers reduced the employment-to-population ratio by 0.37 percentage points and wages by 0.73 per cent during 1990–2007. This amounts to each robot accounting for 6.2 workers. The effects remain substantial, even when they allow for the possibility that robots could lower production costs and enable other industries to increase employment. Dauth et al. (2017) do a similar analysis for Germany, finding that robots have not impacted adversely on total employment, but have done so on German manufacturing employment. They estimate that one additional robot on average replaces two manufacturing jobs, implying that over the period 1994–2014, about 275 000 full-time manufacturing jobs have been lost to robots. These have been compensated for by job gains outside manufacturing – although it is not obvious that this is because of robots. They further find that robots have not directly replaced incumbent workers; rather, they have led to reduced hiring of labour market entrants in more robot-exposed industries. They conclude that robots do not destroy existing manufacturing jobs, but they result in firms creating fewer opportunities for young people. But this is probably a wrong perspective. It is a normal part of a firm's life cycle that workers leave or retire and are then replaced by new workers. That this is not happening represents job destruction. They also find that robotisation has led to increased earnings for high-skilled workers, particularly those in scientific and management positions. However, for low- and medium-skilled workers in manufacturing there are substantial negative impacts. Finally, they conclude that robots raise average productivity and total output, net of labour income, but not average wages, thus linking robots to the decline of the labour income share (Autor et al. 2017).

Graetz and Michaels (2017), using a panel of 17 countries, also find evidence of a negative impact of robots on aggregate employment. In addition, they find evidence that they reduce the hours and the wage bill shares of low- and middle-skilled workers, but with no significant effect on the employment of high-skilled workers. They also find that increased robot use contributed about 0.37 percentage points to annual GDP growth, equating to more than one-tenth of total GDP growth over the period.

On the theoretical side, Hémous and Olson (2014) construct a model where capital complements high-skilled labour, whilst substituting for low-skilled labour, increasing the wages of the former at the expense of the latter. However, slightly contradicting that, Feng and Graetz (2015) conclude that automation should lead to a gradual displacement of middle-skilled workers with a growth in low-skilled and especially high-skilled jobs. As Hémous and Olsen (2014) observe, there are three relevant strands to the literature with respect to this issue. The first relates to Nelson and Phelps' (1966) hypothesis that skilled workers adapt more rapidly to technological change, and that the demand for skilled workers is increased by large technological advances. Secondly is the possibility that capital complements high-skilled labour (Krusell et al., 2000). Thirdly, there is the issue as to whether technology is either high-skill or low-skill labour augmenting (Katz and Murphy, 1992). Sachs et al. (2015) follow a different tack, focusing not on wage inequality so much, but taking a longer view concluding that robots have the potential to decrease the welfare of future generations.

Many, although as we have seen by no means all, of the economists who are concerned with these issues are sceptical on the adverse impact of robots on unemployment per se. Some even dispute that there will be an adverse impact on prosperity. This scepticism may arise because of history, a looking back at previous technological revolutions. In part, too, it may be linked to theory and a belief that technical progress, including AI and robots, destroys some jobs but creates others, often through new products and new markets. There may be short-run adjustment problems, as has often been the case in the past, but not in the long run. To an extent, however, this is a technical issue, rather than an economic one. The key question is the extent to which robots can fully replicate and hence replace human skills and human beings. The position of many economists, for example Autor (2015), is that there is a limit to this process – a limit to the human skills that robots can replicate. Given that these are necessary in many jobs, there are thus limits as to the extent robots can replace human capital in the production function and thus destroy jobs. Autor quotes the example of fitting windscreens on a car production line and fitting replacement windscreens to cars where these have broken. The former robots can do, the latter they cannot. One often meets this type of argument and often it has proved wrong. For example, Autor et al. (2003) argued that navigating a car through city traffic would be difficult for a driverless car. Similarly Levy and Murnane (2004) claimed that it would not be possible to automate driving in traffic. With driverless cars already on the roads and soon to become commercially available, these statements were clearly wrong. The problem is that the argument rests on the current technology, not the technology of the future.

Mokyr et al. (2015) appear to take a different position. They admit the possibility that 'it seems frighteningly plausible that this time will be different', with large sections of the labour market losing jobs. However, in the end they conclude such fears to be misplaced, as the future will see new products that we cannot even begin to guess at, which will combine with new occupations and services that are currently not even imagined. Ultimately too their conclusion depends upon an acceptance of the assumption that there is a limit on the ability of robots to replicate human skills. Indeed. Now clearly there is likely to be such a limit, the question is: what is it? The implicit assumption of many economists is that it is not that great and there will still be plenty of jobs for humans to do. Yet one wonders whether economists are qualified to make this judgment. When we turn to non-economists, more engaged with robotics, we sometimes meet less scepticism. Thus Pratt (2015) argues that technological developments are leading to a revolution in the diversification and applicability of robotics that will result in explosive growth. Thus, with Cloud robotics it is possible that when one robot learns something, other robots will also instantly have access to this knowledge, which obviously leads to a rapid growth of robot competence. Whilst Pratt does not know when the fundamental breakthroughs will be made, he does not doubt that the effects on economic output and human workers are certain to be profound.

Concerns about jobs are not limited to manufacturing jobs, nor repetitive jobs, but extend to a wide range of cognitive jobs (Brynjolfsson and McAfee, 2014). Frey and Osborne (2017) envisage two waves of AI impact. In the first wave, most jobs in transportation and logistics, together with the bulk of office and administrative support workers, and labour in production occupations, are likely to be lost. The first of these is linked to autonomous vehicles (AVs), and AI makes it probable that office and administrative support jobs will go. Some sales jobs require social skills, but cashiers, counter and rental clerks are all at risk. Other non-creative jobs are also at risk. The timing of the second wave will primarily depend on how quickly problems relating to creative and social intelligence can be overcome. In this second wave, occupations least at risk include those in management, business, finance, education, healthcare, the arts and media, engineering, science and the law. However, in contrast with other work and recent trends, they do not predict a decline in middle-income jobs with a growth in low- and high-income jobs. Instead, they foresee a forthcoming attack on low-skill and low-wage jobs.

In terms of the number of jobs at risk, MGI (2013) estimate that sophisticated AI algorithms could replace approximately 140 million knowledge workers across the world. Other estimates vary from about 57 per cent of jobs for Organisation for Economic Co-operation and Development (OECD) countries (World Bank, 2016), 47 per cent of total US employment (Frey and Osborne, 2017) to 55 per cent in Japan (David, 2017) and just 9 per cent in 21 OECD countries (Arntz et al., 2016). The latter differs from the first two studies in taking a task- rather than an occupation-based approach. They argue that some tasks contained in the high-risk occupations are difficult to automate and also resort to the old argument that robots may create new products and services, resulting in occupations we cannot imagine.

5.3 THEORY: THE IMPACT OF ROBOTS ON THE LABOUR MARKET

We illustrate the impact of robots with a diagrammatic analysis that captures much of the algebraic analysis of, for example, Sachs et al. (2015). Figure 5.1 shows three labour markets: a high-skilled labour market where the wage is WH, a medium-skilled labour market and a low-skilled one. Jobs in the medium-skilled labour market are then completely replaced by robots. This is a loss of LM jobs. All of the workers in this market who want to work at a lower wage are then assumed to move to the low-skilled labour market. The shift in the labour supply curve reflects this. The wage falls from WM to WL' for these workers and from WL' to WL'' for the workers originally in the low-skilled labour market. Both lose out, but particularly the medium-skilled workers. This widens inequality as measured by the gap between this wage and WH. The increase in employment in this job market, $LL'' - LL'$, is less than the jobs lost in the medium-skilled job market, LM. But there is no unemployment; workers 'voluntarily' decide to leave the labour market at this lower wage.

The extent to which wages and employment fall depends in part on the elasticities of demand and supply in the medium- and low-skilled job markets, particularly the latter. The less wages fall, the more employment falls and vice versa. But it also depends upon another crucial factor, the ratio $LM/(LM + LL')$, or alternatively the ratio $LM/(LH + LM + LL')$. The first ratio reflects the proportion of non-high-skilled jobs affected by robots. As this proportion approaches 1, so the shift of the curve in Figure 5.1 in the low-skilled job market increases, and wages and employment decline further. The second ratio reflects how important this is for the economy as a whole. If LM is relatively small, then few people will be affected. The analysis

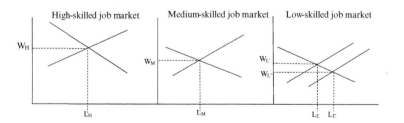

Figure 5.1   The impact of robots on the job market

has assumed that robotisation has no impact on the high-skilled job market. This may not be the case for two reasons. Firstly, it may add to the number of high-skilled jobs that will tend to shift the demand curve to the right and possibly raise wages, although more people may be induced to invest in education. But, secondly, robotisation may impact on this job market too, replacing some jobs just as it has in the medium-skilled job market. We are also assuming that medium-skilled workers do not retrain and move into high-skilled jobs, as the skill gap is too great. An alternative version of the analysis looks, as in Hudson et al. (2017b), at two low-skilled job markets where initially the two sets of wages are identical allowing a focus on the robotisation of low-skilled jobs.

It is possible that, at some levels, increased robot width and productivity may increase the real wage even for the medium-skilled people whose nominal wage has fallen the most, with prices falling still more. There is no unemployment in this model, as is often the case with economic models, due to the assumption that at some positive wage, labour markets clear. But unemployment is a possibility if there is a lower bound to the wage, possibly caused by minimum wage legislation or unemployment benefits. In this case, once the market clearing wage falls below this lower bound, unemployment will result. Unemployment may also result in the short run as the low-skilled sector expands sufficiently to facilitate the employment of the new workers and perhaps for the displaced workers to be retrained for their low-skilled jobs. That is, all the additional workers moving into this labour market may not be employed immediately. In the short run, in the case, we have analysed that it is the medium-skilled workers who are most affected.

In this context it is also the young who will suffer most. Those not destined for the high-skilled market will be in a similar position to the displaced workers from the medium-skilled sector, having to be absorbed into the low-skilled sector at a wage below their expectations, and for some well below. Again, this will take time. In the pre-robot situation all three labour markets would have been absorbing these new workers, who in a steady state would have exactly replaced workers who are exiting the labour force. All industries would be used to this situation and the young would have found work immediately. Now the medium- and low-skilled young will need to wait until the low-skilled sector has expanded sufficiently to absorb both the young and the displaced workers from the medium-skilled sector.

Each year the situation will be exacerbated by some of the young from the previous year still not having found employment.

There is at least one qualification that needs to be made to the analysis. There are many different countries and regions. The region that robotises first may gain at least a short-run competitive advantage over other countries, thus shifting their labour demand curves to the right. There will be an adverse impact on other countries, with some firms potentially closing down. Thus, in the short-run there is likely to be unemployment and downward pressure on wages in these countries. The robotised region or country may thus gain in prosperity at the expense of other competitor regions. Robotisation then represents a positive shock to the former and a negative shock to the latter (Bornhorst and Commander, 2006).

5.4   EMPIRICAL WORK

5.4.1   The Equations We Analyse

In this section we analyse the impact of robots on the labour market using both individual attitudinal data and individual circumstance data. The literature on people's attitudes to robots is linked to a wider literature on the attitudes to technology in general. The deficit model, which has been criticised, suggests that if people are ignorant of the relevant facts in evaluating a new technology, they revert to irrational fears of the unknown (Sturgis and Allum, 2004). To an extent this can be tested by examining the impact of knowledge variables, such as an individual's education, on attitudes. Hudson and Orviska (2011) study people's attitudes to gene therapy, and provide some support for the deficit model in finding that increased individual education is linked to more positive attitudes to this new technology. They also found other socioeconomic variables to be significant, in a manner that reflected the impact of both self-interest and differing degrees of risk aversion. Thus women and older people tended to be less approving, as did those who lived in rural areas and the unemployed. There were also substantial differences between countries; in addition to any differing socioeconomic characteristics, any such differences may partially reflect confidence in the regulatory institutions of different countries. Much of this reflects the pattern of innovation diffusion noted in Chapter 1.

The two attitudes we analyse are, first, perceptions that robots take jobs and, second, that people's own jobs are perceived to be at risk. We argue that the extent to which people will perceive the latter to be the case will depend upon whether their job skills can be easily replicated and the degree of knowledge the individual has about robot capabilities. Hence, we would expect low-skilled manual workers to be more at risk, and thus to perceive this, than more skilled workers or certain non-manual workers. The impact of knowledge variables will depend upon the impact of ignorance on people's views of robot abilities. The deficit model might suggest ignorance leads people to over-exaggerate robot abilities. People's responses to the second question relating to robots taking jobs, will depend upon their knowledge of jobs in general. This will again be linked to knowledge and will be reflected by such knowledge variables as age and education. We also include an individual's occupation, as they are likely to be influenced by their own job that they know best. We also include work robot presence in the region. The more people come into contact with robots, the better informed they should be. Hence, we would expect the responses to these two attitudinal variables to be affected by similar variables, but the knowledge variables to be more important in the latter regression, relating to robots taking jobs, and occupational variables possibly less important.

The individual circumstance variables we analyse relate to the individual being unemployed and their personal prosperity. The control variables will include education, age, gender and location as well as country fixed effects. In the regression relating to prosperity we also include occupational variables. The key variable is regional robotisation, that is, the proportion of the labour force in the NUTS2 region the individual lives in, who work with robots.[1] To reduce the possibility of spurious correlation we also include other regional variables. These comprise regional average unemployment, the proportion of skilled workers and the level of education. The proportion of skilled workers is included to capture industrial structure. When we analyse the attitudinal variables relating to robots we include these same control variables, but add to them regional prosperity. These variables capture both knowledge, reflecting the deficit model, and also individual circumstances. Thus an individual's education and

---

[1]   Subject to the qualification we later make with respect to NUTS2 regions.

occupation are included to reflect the differing probabilities that a specific individual's job can be done by a robot as well as the individual's knowledge. Regional robotisation is important in being purely a knowledge variable in these attitudinal regressions.

5.4.2 The Data

We use the data from the Eurobarometer 87.1 survey carried out in March 2017 in the EU member states, with 27901 respondents participating. We also refer to the Eurobarometer 82.4 survey, which was conducted in November and December 2014. Prosperity was measured by an individual's ability to pay their bills. The description and the coding of all variables can be found in Appendix 5.1 at the end of this chapter. The summary data for selected variables and their distribution according to socioeconomic characteristics are described in Table 5.1. In general, respondents mostly agree with the statement that robots take jobs, a view particularly common amongst the relatively uneducated, those who live in villages, manual workers and women. However, most people also believe that robots could not do their job. There are substantial differences between those in regions of high and low robot density, with the latter being more of the view that robots take jobs. There are also substantial differences between countries. For example, respondents from Portugal, Spain, Cyprus, Greece and Malta are the most convinced that robots take jobs, whilst those in Denmark, Sweden and the Netherlands are less convinced. Comparing this data with that from 2014, perceptions on robots taking jobs have not changed that much. In 2014 the proportion who totally agreed that robots took jobs was 38.6 per cent and in 2017 it was almost the same at 38.3 per cent. But the proportion who thought that it was impossible for robots to do their jobs fell from 59.2 per cent to 52.4 per cent. Clearly people's perceptions of robot abilities are increasing.

5.4.3 Regression Results on Individual Data

All the regressions in this chapter are restricted to regions with more than 50 observations, to reduce the noise in the regional measures. The results of the first set of regressions are shown in Table 5.2. The first column relates to the perceived possibility that robots could do the individual's job. The probability of this decreases with the age

Table 5.1 The summary data for selected variables related to robots

	Could not do my job	Do take jobs		Could not do my job	Do take jobs
All	52.4%	38.3%	Skilled manual	44.9%	42.3%
Young	48.3%	34.9%	Unskilled manual	44.1%	50.8%
Old	54.9%	39.6%	Unemployed	–	45.7%
Highly educated	55.3%	31.4%	Retired	–	41.3%
Medium education	50.6%	37.9%	Students	–	31.7%
Low education	53.1%	48.0%	House person	–	49.1%
Pay bills	54.9%	34.1%	Male	49.9%	35.9%
Village	54.5%	41.8%	Female	54.8%	40.3%
Town	57.8%	37.6%	High robot region	52.4%	34.6%
City	49.8%	35.0%	Low robot region	52.5%	42.8%
Professional–Senior man.	58.0%	28.0%			

Notes: The proportions (1) who did not think that robots could do their job at all and (2) totally agree that robots take jobs.

Source: Based on data derived from Eurobarometer 87.1 (2017).

Table 5.2   *Results of ordered probit regressions on attitudinal variables*

	Coefficients: Robots could do my job	Robots take jobs	Marginal effects: ME1	ME2
Age	-0.00628**	0.00371**	0.00242**	0.00130**
	(6.72)	(5.51)	(6.76)	(5.52)
Male	0.1432**	-0.1294	-0.0551**	-0.0452**
	(6.28)	(8.44)	(6.30)	(8.47)
Log of	-0.06826	-0.513**	0.02625	-0.1793**
education	(0.98)	(12.38)	(0.98)	(12.47)
Village	-0.1259**	0.0855**	0.04841**	0.0299**
	(4.36)	(4.29)	(4.37)	(4.30)
Town	-0.03499	0.0535**	0.01345	0.0187**
	(1.27)	(2.90)	(1.27)	(2.90)
Prosperity	-0.05339**	-0.0788**	0.0205**	-0.0275**
	(2.58)	(5.77)	(2.58)	(5.78)
Regional variables				
Education	-0.00925	0.00722	0.00356	0.002522
	(0.65)	(0.77)	(0.65)	(0.77)
Prosperity	-0.0437	-0.156**	0.0168	-0.0545**
	(0.54)	(2.65)	(0.54)	(2.66)
Unemployment	-0.394*	0.4268**	0.1515*	0.1491**
	(1.98)	(3.30)	(1.98)	(3.30)
Work robots	1.545**	0.47	-0.5941	0.1642
	(6.40)	(2.73)	(-6.43)	(2.73)
Observations	11377	23259		
Log likelihood	-12425	-26724		
χ^2	602.4	3009		
Country effects	Yes	Yes	Yes	Yes
Occupation effects	Yes	Yes	Yes	Yes
Observations	11377	23259		
Log likelihood	-12425	-26724		
χ^2	602.4	3009		

Notes: The regression coefficients were estimated using the ordered probit method; (.) denotes t statistics and * and ** significance at the 5% and 1% levels respectively. Standard errors have been corrected for heteroscedasticity. The regressions are restricted to individuals in regions with a minimum of 50 people. ME1 is the marginal effect relating to the probability that robots could not do their job at all and ME2 the probability that they entirely agree that robots take jobs.

Source: Based on data derived from Eurobarometer 87.1 (2017).

of the respondent. This could be due to a general scepticism about robot ability increasing with age, or the changing nature of their job as people gain experience and perhaps promotion. The only regional variable that was significant in this regression was use of regional robots. This significantly increased such perceptions at the 1 per cent level of significance. Thus, one can conclude that greater knowledge of, and familiarity with, robots increases perceptions of their capabilities. This familiarity argument may also partially explain the significance of the locational variables, village and town. Regional robot interaction terms involving individual education and occupation were not significant at the 5 per cent level and are not included in the regressions. This indicates that the impact of being exposed to robots on people's attitudes to whether robots could do their jobs does not depend on their education or occupation. The marginal effects from this regression are shown in column 3.

The results of the regression relating to robots taking jobs, as shown in column 2 of Table 5.2, are in most respects similar to the previous one. One difference is the much greater significance of the individual education variable. Males are in less agreement with this statement, which is consistent with a literature that finds women to be more averse to new technology than men (Hudson and Orviska, 2011). The results also suggest that a high level of regional unemployment is associated with a perception that robots take jobs. There are differences between the two regressions: people who live in villages and towns tend to be more sceptical that robots could do their jobs, but feel more strongly that robots do take jobs. The same difference characterises the impact of individual age.

The final two columns of Table 5.3 show the regression coefficients associated with the occupational dummy variables in the regressions shown in Table 5.2. The default case is a farmer, hence the zero coefficient. Most coefficients in the 'do my job' regression are positive, indicating that other occupations feel it more likely than farmers that robots could do their jobs. This is particularly the case for the unskilled manual worker who are most likely of all, other things being equal, to feel that robots could do their jobs. Least likely are business proprietors and general management. These coefficients are in approximate agreement with column 2, which shows the average response to the job replication question. For unskilled manual workers this is 1.96, quite near 2, which is a response equating to robots partially being able to do their job. The occupations are

Table 5.3 Occupational data on attitudes to robots

	Sample data on robots to:			Regression coefficients		
	Frequency	Do my job	Taking jobs	Impossible do my job	Do my Job	Taking jobs
Unskilled manual worker	2.84	1.96	3.29	0.44	0.4023	0.1379
Supervisor	0.89	1.88	2.9	0.43	0.3031	-0.2621
Skilled manual worker	8.4	1.86	3.19	0.45	0.2192	0.0633
Employed position, travelling	3.28	1.79	3.2	0.50	0.1550	0.0765
Employed position, at desk	8.3	1.74	2.94	0.49	0.1788	-0.1317
Fisherman	0.02	1.67	3.33	0.50	0.3704	-0.0124
Employed professional	3.03	1.65	2.88	0.55	0.0425	-0.1883
Farmer	0.57	1.63	3.27	0.53	0	0
Employed position, service job	6.89	1.63	3.09	0.57	0.0574	-0.0250
Owner of a shop, craftsmen	2.64	1.61	3.1	0.58	-0.0430	-0.1157
Middle management	6.57	1.58	2.81	0.59	-0.0223	-0.2185
Professional (such as lawyer)	1.25	1.57	2.69	0.60	0.0080	-0.3088
Business proprietors	1.77	1.57	2.9	0.60	-0.0955	-0.1712
General management	1.2	1.57	2.76	0.62	-0.0815	-0.2916
House person	5.74	–	3.3	–	–	0.3469
Student	6.27	–	2.97	–	–	0.1717
Unemployed	6.69	–	3.22	–	–	0.0013
Retired	33.66	–	3.14	–	–	-0.0374

Notes: Column 1 is the proportion of the sample; columns 2 and 3 are the average responses to the two questions; column 4 is the proportion who thought robots could not do their job at all; columns 5 and 6 are the regression coefficients from the first two regressions in Table 5.2.

Source: Based on data derived from Eurobarometer 87.1 (2017).

ordered by this column, with the highest value being first. Column 4 is the proportion who responded that robots would not be at all able to do their jobs. Even for unskilled manual workers it is high. These values give an indication of jobs most at risk, and this tends to be consistent with the literature.

We can use these responses to estimate the number of jobs at risk. We first assume that responses of 'fully' and 'not at all' correspond to proportions of a job at risk of 1 and 0 respectively, and the response of 'partially' corresponds to 10 per cent of the tasks involved being at risk and 'mostly' corresponds to 50 per cent of the tasks at risk. In that case 14.70 per cent of people's jobs are at risk. If we now change the two proportions to 25 per cent and 70 per cent respectively then this increases to 22.45 per cent. The occupations most at risk are as indicated in Table 5.3. Thus, based on these assumptions and using workers' own perceptions, the proportion of jobs at risk from robots in the EU is possibly in the region of 15 per cent to 22 per cent. This is a substantial increase on 2014 where the corresponding figures were 11 per cent and 17 per cent respectively. These changes could represent, even over a short time, the growing abilities of robots, or people's changing perceptions and indeed fears.

We now turn to an analysis of non-attitudinal variables pertaining to being unemployed and prosperity. In the first regression in Table 5.4, the dependent variable is whether the individual was unemployed. The regression was estimated by probit. The critical explanatory variable is regional robot density. This is negatively significant at the 1 per cent level of significance. Thus we can conclude that as robotisation in the region increases, the probability of being unemployed falls. We can thus reject the hypothesis that robots cause unemployment in the region where they are located. Indeed the reverse is the case. Both a robot-education interactive term and regional robot density squared are insignificant at the 5 per cent level of significance and are not included in the results. Thus, the impact of regional robotisation on the probability of an individual being unemployed does not depend upon their education.

The marginal effects are shown in column 2 of Table 5.4. As a specific example, a 45-year-old man, leaving full-time education at 20 with an average level of regional education and skills as in the sample as a whole, in a country such as Ireland, would see the probability of being unemployed fall from 15.0 per cent to 12.5 per cent as regional robotisation increases from 0.05 (5 per cent) to 0.15 (15 per cent).

Table 5.4 Regressions on unemployment and prosperity

	Unemployed	ME	Prosperity	Prosperous	Not prosperous
Age	-0.05462**	-0.00997**	-0.01748**	-0.01533**	0.01985**
	(8.19)	(8.19)	(5.03)	(4.10)	(3.81)
Age^2	0.05145**	0.009386**	0.02943**	0.02855**	-0.02772**
	(6.97)	(6.97)	(8.31)	(7.66)	(5.21)
Male	-0.1073**	-0.01958**	0.09741**	0.127**	-0.02004
	(3.63)	(3.63)	(5.26)	(6.35)	(0.73)
Log of	-1.593**	-0.2906**	1.038**	1.032**	-0.9612**
education	(11.09)	(11.16)	(15.73)	(14.24)	(10.12)
Village	-0.00732	-0.00134	0.02858	0.08223**	0.08688*
	(0.18)	(0.18)	(1.21)	(3.20)	(2.45)
Town	0.05225	0.009534	0.02808	0.0622*	0.04544
	(1.38)	(1.38)	(1.27)	(2.55)	(1.33)
Regional Variables					
Education	-0.00975	-0.00178	-0.04935**	-0.04861**	0.04958**
	(0.47)	(0.47)	(4.16)	(3.76)	(2.72)
Work robots	-5.718**	-1.043**	1.679**	2.118**	0.06324
	(3.73)	(3.73)	(2.66)	(3.04)	(0.07)
Skilled workers	1.020**	0.186**	0.1524	0.1158	-0.3938
	(4.22)	(4.22)	(1.06)	(0.74)	(1.74)
Work robots x	0.2829**	0.05162**	-0.1409**	-0.1433**	0.1139**
individual education	(3.80)	(3.80)	(4.77)	(4.37)	(2.76)

Table 5.4 (cont.)

Unemployment		-0.5531**		-0.7511**	0.3519
		(3.60)		(4.51)	(1.59)
Country effects	Yes	Yes	Yes	Yes	Yes
Observations	13482	22877	13482	22877	22872
Log likelihood	-4506	-16279		-11860	-5761
χ^2	622.2	5118		4419	2101

Notes: The regression coefficients were estimated using the probit method, apart from that for 'prosperity' that was estimated by ordered probit. 'Prosperous' takes a value of one if the individual never or almost never had problems paying bills and not prosperous if they had such problems most of the time; (.) denotes t statistics and * and ** significance at the 5% and 1% levels respectively. The marginal effects (ME) have been calculated for the first model. Standard errors have been corrected for heteroscedasticity in all regressions. The regressions are restricted to individuals in regions with a minimum of 50 people and exclude students.

Source: Based on data derived from Eurobarometer 87.1 (2017).

Individual prosperity has been proxied by a respondent's ability to pay their bills. The fewer problems they have, the higher is their assumed level of individual prosperity. The results from analysing this are shown in column 3. Regional robot density is positively significant at the 5 per cent level of significance. Once more, both a robot density squared term and a robot-education interaction term are insignificant at even the 10 per cent level of significance and are omitted from the regression. However, if we restrict the regression to just males, then the latter is very significant at the 1 per cent level, with the gains from regional robotisation declining with individual level of education. Thus there is evidence that robotisation least benefits well-educated males. The effects of the other control variables are mostly in line with our expectations. Instead of showing the marginal effects, we repeated this regression with two probit ones, the first where the dependent variable equalled one if the individual almost never/never had problems paying bills and the second if they had difficulty paying bills most of the time, that is, the extremes of prosperity. The results are shown in the final two columns. Regional robotisation was significant in the first at the 1 per cent level, but not in the second at even the 10 per cent level. It thus seems that regional robotisation pushes up the prosperity of some, but is less successful in improving the position of the poorest in society. Thus there is no evidence that it is reducing inequality.

5.4.4 Regression Results on Regional Averages

Figure 5.2 plots regional unemployment (Ur) against regional robot density (Rr), that is, the proportion of working people in the region working with robots. Also shown is the fit from a linear regression. It is clearly downward-sloping, indicating that higher regional robot density is linked with lower unemployment. In order to explore this further we extend the regression to include regional education (Er) and the regional proportion of skilled workers (Sr). These other regional variables are again included to reduce the possibility of spurious correlation. The equation is estimated using the Tobit estimator with a lower bound of zero and an upper bound of one. The results are shown below:

$$Ur = 0.370 - 0.491Rr - 0.0107Er + 0.202Sr\,; \quad \text{observations} = 183,$$
$$\chi^2 = 36.86$$

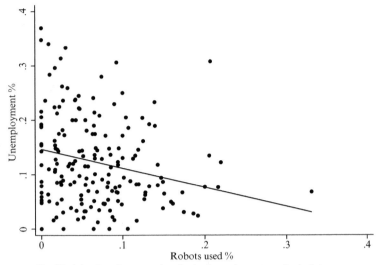

Fitted line is based on a linear regression of regional unemployment on regional robot use

Note: Fitted line is based on a linear regression of regional unemployment on regional robot use.

Figure 5.2   Regional (NUTS2) unemployment and use of robots in the workplace

The coefficients are significant at the 1 per cent level for all variables, apart from education, which is significant at the 5 per cent level.

Figure 5.3 shows similar data, but for the regional average related to paying bills reflecting prosperity (*Pr*). The expanded regression, again estimated by Tobit regression with an upper bound of 3, is shown below. The regional proportion of skilled workers was not significant at the 10 per cent level and has been omitted. Regional robotisation is positively significant at the 1 per cent level of significance and regional education at the 5 per cent level of significance. Thus, once more, there is evidence that robotisation increases prosperity.

$$Pr = 1.725 + 2.135Rr + 0.032Er; \text{ observations} = 183, \chi^2 = 29.55$$

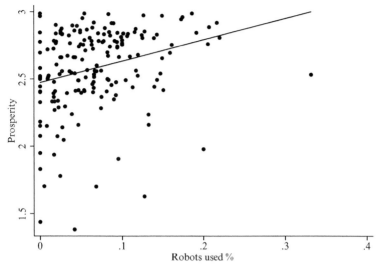

Fitted line is based on a linear regression of regional prosperity on regional robot use

Note: Fitted line is based on a linear regression of regional prosperity on regional robot use.

Figure 5.3   Regional (NUTS2) prosperity and use of robots in the workplace

5.5   CONCLUSIONS

The results suggest that most people believe that robots take jobs and such perceptions tend to increase as regional robotisation increases. Somewhat paradoxically, people are also sceptical about the ability of robots to take their own jobs in the future. When we turn to analyse actual unemployment and prosperity we find that robots have a beneficial impact. Regions with high robot density are characterised by individuals with a lower probability of being unemployed and are more prosperous as reflected by their ability to pay their bills. These empirical results thus confirm the analysis of Hudson et al. (2017b). Hence it would appear that, despite fears to the contrary, robots are benefiting regional labour markets.

However, these results are also consistent with the conclusion that regions with a high incidence of robots gain jobs and prosperity at the expense of other regions. This may be the case if robots enhance a regional employer's comparative advantage in the market, which in turn increases regional demand for labour. But other regions with a lower robot presence lose out. Thus it is possible – though we do not yet have enough data to make a judgment – that the net effect is negative. In this scenario robots impoverish regions that are slow to adopt robotics. We did test for this by including both maximum and average robot usage for all the other regions of a respondent's country. This showed a further favourable impact on unemployment and no impact on prosperity. Hence a high robot presence in one region of a country can benefit the whole country. However, the possibility exists that the job transfer takes place between, rather than within, countries, and looking at our results from the reverse perspective, regions with a low robot usage are experiencing relatively high unemployment. Whether this will prove permanent or a temporary disruptive effect, we do not know.

The people who perceive themselves most at risk in terms of their jobs are skilled and unskilled manual workers and farmers, as these types of people, more than others, both believe robots take jobs and that robots could do their jobs. However, when looking at actual individual prosperity and unemployment, there was no evidence that in practice robotisation adversely affects less well educated people more than highly educated people. Thus the unambiguous policy implication is that, at this point in time, countries should seek to robotise their industries. If they do not and others do, they will suffer. However two qualifications must be made: firstly, that the analysis is based on current technology and we know that this is a technology that is advancing rapidly. Hence it is a moot question as to whether these same conclusions will still be appropriate in two decades' time. Secondly, this is just one piece of research. There is a lot of other research that suggests robots have a more negative impact on the labour market in their own locality.

APPENDIX 5.1

Data are derived from Eurobarometer 82.4 (2014), November–December 2014.

Dependent variables

Unemployed	Coded 1 if respondent is currently unemployed, otherwise 0.
Prosperity	Difficulties to pay bills at the end of the month during the last 12 months. Coded 1 if problems 'Most of the time', 2 if problems 'From time to time', 3 if problems 'Almost never'.
Robots could do my job	Do you think your current job could be done by a robot in the future? Coded 1 if 'Not at all', 2 if 'Partially', 3 if 'Mostly', 4 if 'Entirely'.
Robots take jobs	The extent the respondent agrees with the statement that robots steal peoples' jobs. Coded 1 if 'Totally disagree', 2 if 'Tend to disagree', 3 if 'Tend to agree', 4 if 'Totally agree'.

Independent variables

Age	The age, in years, of the respondent.
Male	The gender of the respondent: Male = 1; Female = 0.
Education	Age at which the respondent finished full-time education.
Village/town	Coded 1 if the respondent lives in a rural area or village/town, otherwise 0.
Occupational variables	Coded 1 if the current occupation is, for example, professional/senior manager, otherwise 0. Occupations also include: middle management, skilled manual, unskilled manual, farmer, house person, unemployed and retired.

Regional (mostly NUTS 2) variables

Education	Regional average of respondents' education
Prosperity	Regional average of prosperity for those in work
Unemployment	Regional average of unemployed respondents as a proportion of those either employed or unemployed , including the self-employed.
Skilled manual	Proportion of those employed, including self-employed, who are skilled manual workers.
Work robots	Proportion of people in the labour force who work with robots.

6. The economic, social and political impact

6.1 INTRODUCTION

Robots and artificial intelligence (AI) will change our lives, and in doing so they will change people, society, the economy and also have an impact on our political systems. We need to distinguish between the impact of work robots, humanoid robots and all the other variations and close relatives. Every day we will interact with robots, in the home, in the car, on public transport, in the workplace, in the supermarket and in the town centre. Robots will keep our towns clean and monitor, and reduce, pollution. They will play a key role in the development of smart cities, particularly with respect to service delivery, waste disposal, transport, infrastructure maintenance, and also in applications related to social care, retail and recreation. They will transform our lives and in many ways hold out a hope for a better and safer future. A future where the old are cared for, society is protected from crime and people in rural areas, particularly in developing countries, have access to good medical care and education, not to mention takeaway pizzas. We have already covered these possibilities in previous chapters. Whilst not ignoring the positive impact of robots, in this chapter we will examine some of the indirect, and often more negative, effects. In the previous chapter we focused on the impact of robots on the labour market. In this chapter we begin by considering the wider economic impact. We then move to examine the impact on both people and society.

6.2 THE WIDER ECONOMIC IMPACT

Robotisation will increasingly reduce the occupations and industries in which people are employed. This in turn will accentuate the divisions in society. It will increase the number of high-paid, high-skilled jobs, at least for a time, and may also increase the number of low-skilled jobs. Middle-skilled jobs and some low-skilled work are likely to become increasingly difficult to find. This will accentuate income inequality. But the real inequality will be between those who have wealth and those who do not have wealth. The wealth owners will become richer through increases in the returns on wealth and increases in the value of wealth. This is already happening. Of course, the high earners are also likely to be the wealth owners, hence they have a double bonus. There are three scenarios to consider.

Scenario one ('The traditionalist scenario')   After some period of adjustment during which unemployment could be high and wages adversely affected, equilibrium will be restored, at a higher level of output and prosperity. New jobs and activities, many of which we can scarcely imagine, will replace ones lost to robots. There will be some changes to the traditional capitalist model, as effectively there will be a new factor of production, robots, which are not just some form of capital, but unemployment and low wages will not be a permanent problem.

Scenario two ('The Sachs et al. (2015), as well as others, scenario')   There is a limit as to the range of jobs robots can do, albeit one that has driven labour into a much reduced range of activities compared to the pre-robot situation. Wages will stabilise at a low level, but unemployment will not be a problem, as the market for labour will clear at a wage above that at which 'voluntary unemployment' sets in.

Scenario three ('The death of capitalism scenario')   There is virtually no limit as such on the range of jobs robots can eventually do. A large, and increasing, proportion of economic activity becomes robotised, leaving labour to compete for jobs in a small, and potentially declining, range of activities. Wages will be driven lower and unemployment will increase substantially, particularly amongst new entrants to the labour market, that is, the young.

Traditionally, capitalism has been based on the capitalist or entrepreneur bringing together labour and capital, and perhaps land too, that is, the factors of production. Labour and capital earn a return, the wage rate and the rate of interest. Traditionally this system has served labour reasonably well, particularly in the long term and with the presence of benign labour laws. Workers have prospered with real wages rising. Firms produce products and services that ultimately consumers – let us simplify, workers – buy with their purchases financed by the wages they earn. But robots have changed the picture: the human input into the production function is being replaced by robots – just another form of capital. Now the model breaks down if labour is excluded from the circular flow of income. Robots do not have purchasing power. They do not go out to restaurants and buy fast cars. They are a form of capital and it is the owners of that capital who benefit. It is tempting to say that they will not spend enough to maintain full employment, but in this world the concept of full employment may become increasingly irrelevant. Rather, the problem is that output is below its full potential, held back by a lack of demand.

In the analysis that follows, we are focused much more on the latter two scenarios and not the first. There are three problems to address, particularly in the final two scenarios, but also to an extent in the transition stage of scenario one. Firstly, there is a problem with purchasing power. Who is to buy the products and services firms are producing? With wages low, there may be a permanent lack of purchasing power adding to the unemployment problems. In order to resolve this, 'a wage' may need to be paid to everyone regardless of whether they work or not, similar to what has been suggested by Reed and Lansley (2016) and others, including to an extent Sachs et al. (2015). This is the universal basic income context we discuss elsewhere. Secondly, with firms no longer so dependent on stocks of workers with specific skills, capital will become even more flexible in where it locates and possibly for how long. Thirdly, inequality between those with jobs, particularly the core number of high-skilled ones, and those without – but also between the owners of capital and others – will grow. Possibly too, it needs exploring, the increasingly high wages paid to the upper echelons of firms and institutions will also grow. The solution to many of the above issues involves levying taxes on heavily robotised firms, but the flexibility with respect to where they locate may cause difficulties. This is because, as soon as a country proposes to levy a tax on them, they may threaten to move. In this respect, a global rather than an individual country solution may be required, if that is feasible.

6.3   THE IMPACT ON POLITICS

Low wages and the potential for unemployment will increase dissatisfaction that will manifest itself in ways that, if not threatening
the democratic system, will at least threaten to change it. The extremists,
promising easy and radical solutions to people's problems will be
increasingly likely to prove attractive, pushing the political systems to
the right, to the left, or both, thus fragmenting and polarising society.
Along with this will come not simply a resurgence of trade barriers,
but also other right/left-wing policies, frequently linked to national-
ism. Such policies are, even without robots, of dubious validity, but
add robots to the picture and it becomes much less likely that the
policies will succeed. If the core of the problem is that automation
is replacing workers' jobs, then relocating firms to the USA will not,
for example, solve America's problems. The question then arises as
to how the disillusioned lower middle classes who voted for extremist
politicians and solutions will react to their failure, once that failure
becomes apparent? Will they relapse into apathy? Migrate to more
prosperous areas? Or will they fall prey to even more opportun-
istic, and extremist, politicians? We would thus expect a drift to
increasingly authoritarian states, marked by civil disorder and states
increasingly disconnected with one another – in many ways, a rever-
sal of the last half of the twentieth century. At such times people
often seek scapegoats, someone to blame for their misfortunes. This
may put minorities at risk, and we can see signs that this is happening
in Europe and North America. But gradually people will realise that
the fault lies with the robots. And at that point, will they once more
follow in the footsteps of the Luddites, trying to destroy them? The
owners of capital will likely respond by increasing security. Crime is
also likely to rise.

Robots will have other impacts on democracy. They will reduce
the need for the individual to interact with their local community.
Modern information technology (IT) facilitates communication with
people in other spatial locations, sometimes in other countries. This
is likely to reduce the commitment of the individual to the local
community and possibly the country. This in turn may reduce their
civic duty, and the ability of communities to function as political
and social entities will be reduced, as will be citizens' willingness
to pay taxes (Orviska and Hudson, 2003). This may also impact on
an individual's commitment to the country, although the financial

dependence of many individuals on the state may have the opposite effect.

Finally, AI poses a rather different and more direct threat to democracy. In the future journalists may write a basic story that will then be adapted by AI to suit the views of the recipient, which is learnt from their previous digital and shopping behaviour. The threat is there that many people will only be presented with news that reinforces their views, which will further polarise society. There is also the threat, currently very real, that malign forces may use AI to sow discord in the country and promote the cause of extremist politicians.

6.4 THE IMPACT ON PEOPLE

Robots replace people. How will people react? Throughout history humans have sought satisfaction from work. What will happen to them as that is taken away from them? Not just work in the workplace, but also work outside the workplace. Eventually, they will no longer drive cars, do much of the housework, or care for elderly relatives. Theoretically they will have more time for leisure. They can spend more time pursuing their hobbies and interests. But what will the impact of this be on their personality, and will it increase happiness and wellbeing? Robots will also become more and more like companions, but unlike human companions they can be programmed to never argue or disagree with you. What will be the impact of this on people's personalities and will it lead to disengagement from other humans?

Social capital results from regular social meetings with other people. Thus if such social interaction is reduced as a consequence of robots, then over time there will be a reduction in social capital, particularly 'bonding social capital', associated with closed networks such as family and close friends. 'Bridging social capital' – that is, social capital that cuts across different networks – is often formed in the workplace (Coffé and Geys, 2007). Both are important in helping the individual in different ways. But it is possible that they may be replaced or at least augmented by 'robot social capital', that is, the network, via the Cloud, that a person's robots are in contact with. IT will help the individual solve problems, deal with local bureaucracy and firms, and provide information. In this way robots will reduce social capital, but also reduce the need for social capital. Robots will be people's companions and help them solve life's problems. People will become more and more dependent on robots. Little, if anything, has been written on robot social capital, but we believe it will become increasingly important. However, there are also forces pulling in the other direction. People will have more time for their hobbies and leisure activities, and if these entail meeting other people, for example at the golf club or the allotment, social capital will be enhanced. For many people, too, social capital is expanding beyond the immediate locality and becoming more global. Their associations are often online. That is of use in helping people pursue their leisure activities, but of less use in helping them live in their community and dealing with the everyday problems of living.

Human beings maximise utility subject to an income constraint. Entering into the utility function are their income and leisure time, their own abilities, the environment they live in and the other people they interact with. Robots will change all of this. Firstly, as already discussed, robots will impact on incomes, sometimes positively, but often, unless governments take corrective actions, negatively. This will expand or contract their options. People may well have more leisure time, although for an individual, income and leisure time may change in opposite directions. People's abilities may also change because robots can provide humans with more options, abilities and knowledge. Hence, each individual can have their own robots, which will clean the house, do the garden, inspect the roof, clean the gutters, and protect them. But their abilities will change more directly because of robot exoskeletons that enhance human physical power. In this part of the robot story, robots do not replace people, rather they enhance their abilities. Quite apart from the old and infirm being more mobile, the fit and healthy will be able to walk further and longer. This will have direct implications on people, but implications to which relatively little thought has been given. For example, will exoskeletons make people fitter or less fit? If they use this extra ability to walk further and use their own muscles more, they will become fitter. But if they come to rely more on the exoskeleton and less on their own muscles, the reverse may be the case and we can expect to see obesity increase. This may also happen as robots do more and more of the jobs in the home and at work.

But robot exoskeletons will also enhance people's fighting abilities, indeed this is why so much of their development is focused on the military. This is likely to increase injuries and deaths from fights that occur in normal life, for example in town centres on a Saturday night. The abilities of criminals will also be enhanced, including violent criminals. Robotics will also change the nature of the criminal. Through much of history, strength and ruthlessness were key attributes for a criminal. Today scientific knowledge is becoming more important than at least the former. The computer or robotics expert is able to use opportunities that others cannot and, as Sharkey et al. (2010) comment, this opens the way for a new breed of criminal, for example teenagers working from their rooms. Exoskeletons are only half – the physical half – of the story when it comes to enhancing people's abilities. The other half relates to increasing their mental abilities through brain computer interaction (BCI). As discussed earlier, this tends to work in one of two ways. Firstly, via an electroencephalogram (EEG) cap worn on the head that detects neural activity using electrodes placed onto the scalp. Alternatively they can involve a device planted directly into the brain. Such implants could technically change people into cyborgs. In this case one wonders about the future of, and need for, formal education. There is also the risk that such implants could be hacked and this could be used both to control an individual's actions and to extract information.

The impact technology can have on society is illustrated with the smartphone, where of course bots play an integral part. Turkle (2011) laments that when businessmen used to talk, for example in taxis on the way to the airport, they now do emails. Previously these conversations were when global teams solidified relationships and refined their ideas. The same goes for others, particularly the young. Turkle argues that this makes the young confident in the use of technology, but leads to more artificial relationships. Sarwar and Soomro (2013) discuss further problems. In Korea it is estimated that 8.4 per cent of smartphone users are addicted to their use. Compulsive communicating has been recognised as a serious psychological problem. Sarwar and Soomro also argue that in smartphones taking over functions that a human brain can and should perform perfectly well, the brain does not get the exercise it needs. This can impair people's mental abilities. Bots are, of course, linked to smartphones, but AI takes this further. Instead of using a map to guide one's directions, or choosing a restaurant to dine in, people use an AI device that gives instructions. Indeed, AI can now even book your table for you.

Hence people risk becoming both physically and mentally less fit and able through over-dependence upon robot technology.

6.5   THE SINGULARITY AND RELATED HYPOTHESES

Throughout the book we have dealt with revolutionary developments, but ones that seem reasonably likely to occur, even if they fundamentally change our environment. We now move into a much more uncertain scenario and consider that AI may either totally replace or fundamentally change human beings to the point where they are not recognisably human anymore. Fantastical as it may seem, there is a considerable body of serious scientific literature on this, dating back to the early 1950s, and we must take it seriously.

The singularity hypothesis, of which there are several versions, is that accelerated technological change leads to a situation where humanity in its current form is challenged. In one version the emergence of artificial super-intelligent agents are the 'singular' outcome of this process. These agents, or software-based intelligent minds, engage in a continuous and accelerating process of self-improvement, increasing machine intelligence beyond that of any human being. This thus fulfils Turing's observation, made in a lecture in 1951, that at some stage we should have to expect the machines to take control (Turing, 1951). The implications of this are profound and indeed potentially dire for humanity as we know it (Sandberg, 2010). There is even a fear that self-motivated robots and AI will seek more and more resources for themselves, as they seek to complete their missions, leaving less for people (Korinek, 2018) in a Malthusian race we will lose. An alternative view, marginally more favourable to human beings, is that technology amplifies human physical and cognitive capabilities to such an extent that it can overcome all existing human limitations, in the process living almost for ever (Kurzweil, 2005). Korinek (2018) has similarly argued that robotisation offers the possibility for some humans to enhance their physical and mental abilities. This may help humans compete with robots, although at the price of considerably changing human beings. But this also opens up the way to far greater inequality than we have ever seen before, with the richer people more able to purchase these enhancements. In future the elite may not just have more resources and power, they themselves may also have greater intellectual and physical power. If this were the case it could well, at the very least, substantially slow down social mobility.

This is a controversial area. It is argued that singularity depends upon technological acceleration, which in itself is controversial. This may not be necessary, however; there may simply be a steady advance in technology to the point at which robots outperform and out-think humans by a sufficiently great amount. With respect to the other side of the argument, robotic developments will enhance human abilities. But whether the sum total of this will be sufficient to progress humans to the next stage of development is less clear, nor indeed whether AI will replace humans. Indeed the whole approach has been called speculative and even more a religious notion than a scientific one (Eden et al., 2012). My own view is one of concerned scepticism, certainly on the more apocalyptic possibilities. But the concerns are genuine. The consequences are so serious that it would be foolish to simply dismiss them as being unlikely, or too far in the distant future to worry about.

6.6   THE IMPACT ON SOCIETY

Robots will change society in many ways. Firstly, they have the potential to significantly improve public services with respect to education, health, security, transport, indeed every single aspect of the public sector. But the shift of emphasis from workers to robots may, as we have already said, pose problems for collecting tax revenue. If this is the case, and if governments need also to fund low-paid and unemployed workers with greater transfer payments, then there could be a squeeze on the resources available for the public sector and this will tend to push down the quality and quantity of public services. On the plus side, however, the increased ability for AI to check transactions may well reduce tax evasion, although it may also facilitate money laundering. In changing the economy, robots and AI will also impact upon human geography. The rapid decline of the high street as a place where shops can be found is in part linked to the growth of online shopping and thus AI, and shopping malls may well quickly follow. Both need to evolve into providing services other than retail ones. It is an illustration of how quickly robots can combine with the invisible hand to change the economy and society.

It was discussed earlier how robots can be used to enhance security and policing. But there is another side to this possibility, as was again mentioned earlier. Robot exoskeletons will probably increase or at least change the nature of crime. This is also true of robots more generally. Sharkey et al. (2010) argue that, in the future, perhaps a combination of ground robot assailants and aerial look-outs could be involved in bank robberies, street hold-ups and the theft of high-value delivery trucks. This is assisting already existing criminal activity. But robots also open up the way for new forms of crime in traditional areas and for totally new forms of crime. In the former category we have the possibility that miniaturisation may make possible the sending of machines through letter boxes or the opening of windows to neutralise alarms, steal keys or even small-size valuables such as jewellery. They could also lie hidden in the house, secretly filming and recording activities, which may then be used for blackmail. Underwater autonomous robots can already be used to transport drugs. Autonomous robots in general may also remove the criminal from the crime. In terms of a totally new form of crime, we have the criminal hacking and thence freezing of computers and demanding a ransom to allow them to work. In future the same possibility exists for autonomous vehicles (AVs), indeed anything that is connected to the Internet. Whilst not traditional criminal activity, AVs and robots could also be used by terrorists in attacks using not only explosives, but also guns and lasers.

People are likely to react in various ways to a technologically driven environment from which they have become alienated. It is possible that they will follow the Luddites in trying to destroy the robots. But, there may be simply too many to destroy and they may also be too well protected. It is more likely that some will take a more peaceful turn, and will collectively try to lead simpler lives as the Amish have done in America. Others may react by trying to change, through violent acts, not robots directly, but the society that has seen them progress, as Baader Meinhof tried to do in Germany in the latter half of the twentieth century, in part because they had become alienated from that society and its values.

Further into the future, the possibilities become even more frightening, as we move to a world, somewhat late in arriving, but similar to George Orwell's (1949) novel *Nineteen Eighty-Four*. It is possible that the state will continually listen in on people's conversations in their home, at their work, whilst driving, whilst walking in the park, anywhere where there is a connection to the Internet. The policing of this state will then be done partly by humans, partly by androids and partly by other robots.

6.7 AN INCREASINGLY ELDERLY POPULATION

With an aging population in many countries, societies are under pressure to care for the elderly. In days gone by, the children would undertake that role, as they would live close to their parents. In today's society this is often not the case, and even when it is, caring for an elderly relative, often with poor health, can place an enormous burden on the younger carer. Robots offer the promise of help with all of this. They offer the hope of cheap care for the elderly provided by the public sector and also the possibility of the robot providing help and support for the younger carer. They will also provide the elderly, particularly those who have difficulty in driving, with increased mobility through driverless cars. Exoskeletons can also help those with mobility problems.

But there are potential indirect effects. Let us assume that people have children for three reasons: (1) as a source of income insurance in old age; (2) to help with caring in old age; and (3) companionship. Robots undermine all of these. Firstly, they may reduce the income the next generation will earn whilst increasing the income that can be earned by investing in capital. Hence the rational individual will invest more in capital, and less in people by having children. Secondly, robots can provide care in old age, both health care and doing household jobs, with children thus again becoming less relevant. Finally, robots may also provide companionship. For all of these reasons we would anticipate that the development of robots will reduce the birth rate, leading to an increasingly elderly and, despite medical developments, possibly smaller population. It is also possible that the marriage rate will decline with robots. Robots will provide companionship of all kinds and that may be a substitute for marital relations. Hence the impact of robots may well be to reduce population size, increase the proportion of the elderly, increase the proportion of childless couples and also the number of people who do not get married and live alone – with their robot. This all has implications for the housing market. There will be a demand for smaller houses and flats. This, in part, is what the data show has been the trend for some years. Robots may have had some limited impact on this, although there will be other factors at work too. But the argument we are making is that robots will exacerbate these trends in future years.

6.8   THE IMPACT ON GLOBALISATION, DEVELOPING NATIONS AND ECONOMIC CONVERGENCE

In recent years the global economy has been characterised by firms offshoring, that is, moving their activities to cheaper locations, often developing countries. The rationale is the cheaper labour that is based there. But if robot workers come to replace human ones, then this cost advantage will decline, if not disappear. There is then a risk that the jobs will move back to the more advanced countries. It is not just that there is reshoring of jobs to the advanced countries, but also to the more advanced ones in the developing world such as China and Korea (Maloney and Molina, 2016). With the emphasis on robot sophistication rather than wage costs, the less technologically developed countries may be at a permanent disadvantage relative to other countries. We saw earlier that robots may not be causing overall job losses in countries with a high robot density, but in countries with a low robot density. In these advanced countries the problem may be more one of rising inequality, a problem that can be solved, as always, by government redistribution. But how to redistribute wealth from the rich countries to the poor ones? More aid? Having said all that, increased use of robots may well increase the absolute, rather than the relative, level of prosperity and the availability of food in developing countries.

6.9   THE IMPACT ON RURAL AND PERIPHERAL AREAS

If robots will be most widely used in large towns in developed countries, their potential impact is arguably much greater in remote, rural communities and developing countries. Drones open up the possibility of connecting remote areas with the products and services of large towns, including for example delivery of medical products. Indeed it is already beginning to happen. In Rwanda, blood and plasma are being delivered by autonomous drones to remote rural areas. Deliveries that previously took hours and even days can now arrive within minutes. The same is happening with respect to remote areas in the USA. Drones also have the potential to make deliveries from retail outlets as well as takeaway food outlets, and again it is the rural areas who will benefit substantially from this. Remote medical and educational services will also benefit from robots. Telepresence robots enable visual and audio communication between a remote doctor and a patient and local medical staff. Robot sensors will enable navigation around the hospital. Robots also open the possibility of surgical operations being done in small hospitals by a surgeon many miles away and, in the future, possibly by a robot themselves. Similarly with education, students will be able to 'attend' and participate in schools many miles away. Robots are also being used to help in the care of dementia sufferers in remote communities. A relative or carer can move the robot around the person's home, checking everything is right and also engage in audio and visual communication. These relatively remote communities can also benefit from AVs, caring robots, house robots and security robots, just as any other community can. All of this makes the rural area, in both developing and developed countries, a more attractive place to live. The impact of robots on jobs will be similar to that in urban areas, in that robots will be used in agriculture and this will reduce the demand for farm workers whilst increasing the output of the farms. This will also be the case for other rural industries. The increased viability of home working, facilitated not just by the Internet, but also by telerobots will further increase the viability of rural areas.

6.10   THE KNOWN AND UNKNOWN UNKNOWNS

So far we have mainly dealt with what we know and possibilities about which we are a little uncertain. But as economists tend to say with respect to jobs, there will be implications of robots that we cannot yet imagine. We cannot imagine them for two reasons. Firstly, we cannot imagine the full range of tasks and uses robots will be applied to, nor can we imagine the full range of different types of

robots. What, for example, is the range and the future for nanorobots? But, secondly, we cannot anticipate all the economic and social
impacts of even known robots such as with AVs. For this reason we
must expect the unexpected and be ready to respond, in regulatory
terms, when the unexpected happens, or is about to happen.

7.   People's hopes and fears

7.1   INTRODUCTION

We have examined people's attitudes to robots in a specific labour market context. In this chapter we explore the public's attitudes to robots in specific contexts outside the workplace. Basically we have seen that robots in the home and in society offer both opportunities and benefits as well as risks. How do the public balance these two out? The attitudes we focus on are: (1) caring for the elderly; (2) education; (3) surgery; and (4) driverless cars and trucks. These are all areas that, although they have job implications, also have implications for people's quality of life. In addition we analyse robots for use in dangerous situations and also general attitudes to robots. This work complements only a relatively small literature on attitudes to robots, but there is more on attitudes to technology per se. The analysis is unusual in analysing attitudes to robots across a wide range of uses and allows us to detect factors that consistently influence all, or at least most, attitudes. We begin by examining the relevant literature, focusing on caring robots and driverless cars.

7.2   THE LITERATURE

There has not been that much work done on attitudes to robots, although recently several papers by Marta Orviska, Jan Hunady and John Hudson have added to this literature (Hudson et al., 2017a; Hudson et al., 2018; Orviska et al. 2018). The specific contribution of the analysis in this chapter is that it is the first to look at, and hence compare, attitudes to robots across a range of different uses using regression analysis. The work that has previously been done suggests that a number of factors influence such attitudes and for a number of different reasons. Included in these are culture, experience of robots and the type of robot (Nomura et al., 2009). Nomura et al. also discovered that younger people were found to have more favourable attitudes than older people, a result that has also been reached by others, for example Orviska et al. (2018). There is a degree of hostility to robots, and Coeckelbergh et al. (2016), using Eurobarometer data, found that 60 per cent and 34 per cent of EU citizens want robots to be banned in care and education respectively. However, this does differ from country to country and depends upon the context in which robots are used, hence they also found that, in Romania, Holland, Belgium and England, there was considerable support for the use of social robots in treating children with Autism Spectrum Disorders. Moon et al. (2012) also found majority support for the use of robots in caring for the elderly, but this was subject to the qualification that robots should assist, and not replace, human care. Taipale et al. (2015), using 2012 Eurobarometer data, concluded that the use of caring robots was supported more by both pensioners and students. Often, however, as already noted, older people are more hostile to robot care for the elderly (Broadbent et al., 2009). There is also evidence that women are more hostile than men and that increased education tends to make people more favourable towards robots (Hudson et al., 2017a). Finally, people tend to be more in favour in cities and large towns than in less densely populated areas (Taipale et al., 2015).

Turning to autonomous vehicles (AVs), Kyriakidis et al. (2015) use an Internet-based survey to analyse 5000 responses from 109 countries. They found 22 per cent of the respondents did not want to pay for an AV, whilst 33 per cent thought it would be highly enjoyable, although more felt manual driving provided greater enjoyment. Other surveys tended to find favourable attitudes to driverless cars in California (Howard and Dai, 2014) and also in a sample covering the US, the UK and Australia (Schoettle and Sivak, 2014). Hudson et al. (2018) and Orviska et al. (2018) find a combination of socioeconomic variables and a degree of robotphobia to impact on attitudes to AVs and also robot surgery. Much of the work on people's specific fears with respect to driverless cars has been done by Schoettle and Sivak (2014), Kyriakidis et al. (2015) and Bansal et al. (2016). This work suggests that concerns relate to system failure, the problems of interacting with other, conventional, vehicles, affordability, software hacking, legal issues, privacy and safety. The perceived benefits relate to fewer crashes and better fuel economy (Bansal et al., 2016). Schoettle and Sivak (2014) also found that higher-income, technology-literate males living in urban areas had a greater willingness to pay. Experience of crashes also had a positive effect on attitudes. Men are often found to be more favourable than women (Payre et al., 2014; Bansal et al., 2016; Hudson et al., 2018). Bansal et al. (2016) also found that support increased with income and technological literacy and was higher for urban dwellers. All of this does suggest that when people are evaluating a specific robot use, they are considering the advantages and disadvantages of that robot use.

7.3   BACKGROUND

From an economics perspective, an individual will be in favour of robots if the perceived gains outweigh the costs. Focusing on elderly care, for younger people the former will relate to both lower taxes from the nationally funded social service system and possibly having to pay less themselves in providing for the care of elderly relatives. Potential gains may also relate to improved care for the elderly, if the robot can either provide new services that are not currently offered to most people or they improve on currently offered services. For example, robots may facilitate independent living amongst the elderly (Bedaf et al., 2015). However if robot care is imperfect, there is potential for an adverse impact on the wellbeing of the elderly. This cost will mainly be borne by the elderly themselves, but also to an extent by their younger relatives. For young people, any direct costs related to robot care for themselves are likely to be some years in the future and hence will be discounted. In addition, robot technology can be expected to improve, so that at least some of the problems that currently exist with caring robots will have been solved by the time the young may need robot care. For older people, however, the potential costs are either in the present or possibly in the near future, whilst the tax gains may not be so great if they are not working. But if the older person needs to pay for their own care, then the direct costs are also very relevant. On balance, though, we would expect approval for robot care to decline with age, if there are reasons to be concerned about the quality of such care. In addition, attitudes to robot care will also in part depend upon the current state of elderly care. This is primarily based on human contact and interaction. In many countries there are concerns about this, and stories about mistreatment in care homes frequently figure in the news. These concerns may be more prominent in poorer countries where the standard of care provision is more cost constrained.

A similar story will apply to other robots, with a balancing of risks and gains as they pertain to the individual and the type of robot. For each type of robot, the risks and gains will differ. In many cases, the poor and those in rural areas and small towns will be less likely to gain, simply because they will not be able to afford the robot, or robot-related care, or because the infrastructure is unlikely to be in place locally for them to be able to access the robot. However, as we argued earlier, in the longer term it is the remoter areas that may well gain the most from robots. In addition, women tend to be more risk-averse than men (Hudson and Orviska, 2011) and, hence, perhaps to emphasise the risks more. In most but not all cases, this may lead women to be less favourable towards robots than men. The individual is also likely to be influenced by their attitudes to any one type of robot and by their attitudes to robots per se, as well as to technology in general. Thus, people who fear their jobs are at risk from robots may well be less favourable on all dimensions. There is a considerable literature on general attitudes to robots, which does find attitudes to be less favourable amongst women, the older population, the less educated, people in low-population-density areas and the less well-off. Much of this is also true for technology in general. The deficit model of public attitudes towards science, which was referred to in Chapter 5, suggests that ignorance of the relevant facts leads people to revert to irrational fears of the unknown. We argued in Chapter 5 that such knowledge can be linked to education, which would suggest that an irrational hostility would increase as the individual's level of education declines. Combining all of this together, we can link attitudes to robots to an individual's age, gender, income, location and education. We would also expect there to be differences between countries. In addition, increased familiarisation with robots, which increases awareness of what they can do, can be expected to impact, perhaps positively, on attitudes to robots.

7.4   PEOPLE'S ATTITUDES TO ROBOTS

We use the data from the same Eurobarometer surveys analysed in chapter 5. The independent variables are as defined in Chapter 5. The dependent variables relate to people's attitudes to robot surgery, educational robots, caring robots, AV cars and trucks, robots for use in dangerous situations and a variable relating to robots in general. These are defined in Appendix 7.1 at the end of this chapter. The robot approval data relate to both 2014 and 2017 and thus allow us to focus on how attitudes are evolving over time, albeit a short period of time. The other data relate to 2014, as this survey had information on a wider range of robot uses than the 2017 one. The summary data for selected variables and their distribution according to socioeconomic characteristics is described in Table 7.1. On balance, people are more approving than disapproving of the concept of robots in general terms. This enthusiasm is not spread evenly across the population, and is less prevalent amongst the old than the young, the relatively uneducated rather than the well educated, those who live in low population density areas, unskilled manual workers rather than pro-fessional people or managers, and females rather than males. Hence there is a degree of concern amongst the public about their use, and it is concern that is increasing over time with robot approval declin-ing slightly between 2014 and 2017. People are most enthusiastic about their use in dangerous situations. But when it comes to more specific uses when robots come into contact with people, there is less enthusiasm. This is particularly the case for surgery and driverless cars, but also driverless trucks and in caring situations. There is more approval for their use in education. The socioeconomic variations noted earlier remain evident for most of these options. But there are differences. Thus the gap in attitudes between the young and the old is much less for robots used in education than robotic elderly care and driverless cars.

However the views are not normally distributed, as can be seen in Figure 7.1. This shows the kernel densities,[1] with a bandwidth of one, for attitudes of young and older people respectively to robotic care for education. The dividing age between young and old is 40 years. It can be clearly seen that there is a large peak at the first and most negative view, and a slightly smaller peak at the last and most positive answer. The kernel density for younger people is substantially below that of older people at the first peak and then above it for responses of 6 or more. Clearly younger people are consistently more positive than older people, although even for older people a substantial minority

---

[1] May be thought of as a continuous histogram.

Table 7.1 The summary data for selected variables

	Approval 2017	Approval 2014	Dangerous situations	Surgical	Education	Care for elderly	AV cars	AV trucks
All	6.75	6.87	8.39	3.90	5.42	4.43	3.65	4.32
Young (<40 years)	7.07	7.12	8.39	3.98	5.68	4.85	4.12	4.77
Old (≥40 years)	6.61	6.75	8.39	3.88	5.29	4.22	3.42	4.10
Highly educated	7.34	7.46	8.76	4.79	6.22	4.77	4.49	5.25
Medium educated	6.85	6.98	8.46	4.00	5.55	4.60	3.74	4.45
Poorly educated	6.17	6.36	8.06	3.24	4.76	4.00	3.01	3.56
Prosperous	6.91	7.06	8.58	4.16	5.71	4.54	3.79	4.46
Village	6.58	6.73	8.32	3.76	5.21	4.24	3.46	4.09
Town	6.79	6.88	8.4	3.86	5.42	4.4	3.56	4.19
City	6.90	7.03	8.47	4.14	5.68	4.7	3.98	4.77
Professional/manager	7.45	7.49	8.71	4.91	6.31	5.05	4.79	5.60
Skilled manual	6.72	6.86	8.24	3.71	5.32	4.57	3.70	4.32
Unskilled manual	6.27	6.40	8.12	3.05	4.71	4.05	3.19	3.67
Unemployed	6.52	6.56	8.17	3.38	5.07	4.3	3.46	4.04
Retired	6.43	6.64	8.41	3.75	5.08	4.09	3.11	3.78
Student	7.36	7.48	8.51	4.16	6.21	5.39	4.57	5.19
House person	6.18	6.26	8.03	3.12	4.49	4.03	2.98	3.49
Male	6.99	7.14	8.53	4.41	5.76	4.75	4.17	4.83
Female	6.55	6.65	8.28	3.49	5.14	4.17	3.21	3.89

High robot density	6.88	7.03	8.48	4.20	5.70	4.51	3.78	4.42
Low robot density	6.59	6.67	8.27	3.52	5.06	4.34	3.48	4.18
Driver	6.83	7.06	8.39	4.18	5.41	4.54	3.96	4.53

Notes: The values are the means of the variables for selected categories. The coding for the last five columns ranges from 1 (totally uncomfortable with this) to 10 (totally comfortable). For the first three, the data has been adjusted to be on a scale of 1 (least favourable to robots) to 10 (most favourable to robots). Hence a response of 5.42 is in all cases slightly more favourable than unfavourable. Prosperous is defined as those who never or almost never have problems paying bills.

Source: Derived from Eurobarometer surveys 87.1 (2017) (column 1) and 82.4 (2014).

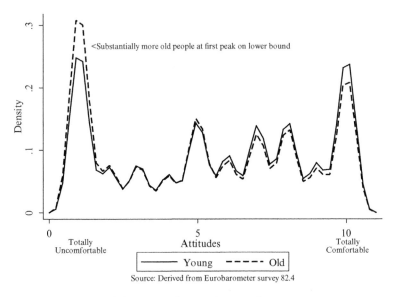

Figure 7.1   Kernel densities of attitudes to robots

favourably view the possibility of robots in education. This is very similar to people's attitudes to robot surgery (Orviska et al., 2018).

Figure 7.2 further illustrates the relationship between age and attitudes to educational robots. It plots the average response for each age against that age for several different countries. Superimposed on the plots is a line of best fit. As can clearly be seen it slopes steadily downwards for most countries and is flat for others. The relationship with age looks to be largely linear, with people tending to become more hostile to robotic care as they age. However there are substantial differences between countries both in average attitudes and the slope of the line.

The most hostile overall are Cyprus, Greece, Hungary and Malta. The most favourable are Denmark, Sweden, Bulgaria and the Netherlands. There is a degree of consistency between the different rankings for different robot uses, but there are also differences. Thus, for surgery, the least favourable countries are Malta, Lithuania, Latvia and Croatia, whereas Lithuania is one of the most favourable for driverless cars. Of course these differences between countries could reflect either different basic attitudes, possibly linked

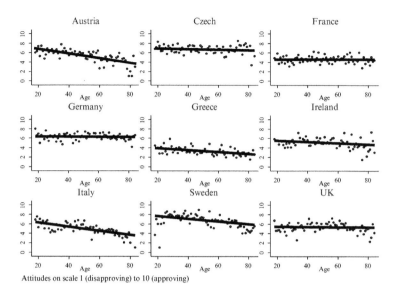

Attitudes on scale 1 (disapproving) to 10 (approving)

Figure 7.2   Attitudes to educational robots across selected countries

to institutional factors, for example relating to the current state of elderly care, or they could reflect differences between countries with respect to their socioeconomic characteristics. In order to differentiate between these different possibilities, we turn in the next section to multiple regression analysis.

7.5   REGRESSION RESULTS ON INDIVIDUAL RESPONSES

In this section we estimate the basic relationships that explain individual attitudes to robots in different, non-work contexts. The regressions were estimated by ordered probit and are shown in Table 7.2. In the first two columns we show the results relating to general robot attitudes in 2017 and 2014. There is a large degree of consistency in the two sets of results. Approval, at the 1% level, significantly declines non-linearly with age, with just the squared age term being significant. This means that people become increasingly hostile as they age. Women and the less well-educated tend to be less approving, as do those who live outside large towns or cities, particularly in

Table 7.2 The regression results

	Approve 2017	Approve 2014	Dangerous situations	Surgical	Education	Care for elderly	AV cars	AV trucks
Age	0.0001	0.00099	-0.00671**	0.00312	0.0112**	-0.00614**	-0.00702**	-0.0057**
	(0.04)	(0.34)	(2.69)	(1.37)	(4.72)	(10.73)	(11.88)	(9.84)
Age²	-0.00817**	-0.00825**	0.00638*	-0.00822**	-0.0129**			
	(2.91)	(2.84)	(2.52)	(3.52)	(5.24)			
Male	0.1722**	0.2154**	0.00490	0.2100**	0.3037**	0.2094**	0.3454**	0.3185**
	(9.72)	(11.98)	(0.32)	(15.08)	(21.12)	(14.97)	(23.87)	(22.39)
Village	-0.0655**	-0.0959**	0.0324	-0.0628**	-0.0949**	-0.1097**	-0.0953**	-0.123**
	(2.87)	(4.03)	(1.58)	(3.43)	(4.97)	(5.89)	(4.96)	(6.52)
Town	-0.0131	-0.0624**	-0.00299	-0.0693**	-0.0638**	-0.0771**	-0.0781**	-0.1013**
	(0.60)	(2.79)	(0.16)	(4.05)	(3.61)	(4.53)	(4.43)	(5.87)
Log of education	0.9031**	0.6929**	-0.1414**	0.4802**	0.6033**	0.2781**	0.6221**	0.6506**
	(18.78)	(14.16)	(3.39)	(12.98)	(15.73)	(7.47)	(16.16)	(17.27)
Professional/ senior man.	0.1134**	0.1320**	0.1001**	0.0896**	0.1376**	0.0828**	0.1732**	0.1916**
	(2.74)	(3.17)	(3.10)	(3.13)	(4.78)	(2.94)	(5.99)	(6.59)
Unemployed	-0.110**	-0.1576**	-0.0406	-0.0651*	-0.1026**	-0.0415	-0.0769**	-0.0933**
	(3.00)	(4.74)	(1.37)	(2.38)	(3.68)	(1.52)	(2.73)	(3.39)
Retired	-0.0381	0.0204	-0.0312	-0.0306	-0.0089	0.0115	-0.0883**	-0.0830**
	(1.27)	(0.68)	(1.17)	(1.26)	(0.34)	(0.48)	(3.52)	(3.41)
Middle	0.0504	0.1715**	0.0535	0.0893**	0.0205	0.0240	0.0591*	0.0814**

	(1)	(2)	(3)	(4)	(5)	(6)	(7)	(8)
management	(1.35)	(4.03)	(1.67)	(3.17)	(0.72)	(0.87)	(2.09)	(2.89)
Skilled manual	-0.0929**	-0.04879	0.0108	-0.0892**	-0.0911**	-0.0550*	-0.0782**	-0.0965**
	(2.73)	(1.43)	(0.36)	(3.30)	(3.16)	(2.03)	(2.80)	(3.49)
Unskilled manual	-0.2509**	-0.1561**	0.057	-0.1951**	-0.1765**	-0.0742	-0.1216**	-0.1456**
	(5.06)	(2.98)	(1.12)	(4.45)	(3.53)	(1.67)	(2.62)	(3.23)
Driver	-0.0200	0.0187	-0.0255	-0.1051**	-0.0120	-0.0839*	-0.05444	-0.1065**
	(0.40)	(0.36)	(0.59)	(2.77)	(0.30)	(2.30)	(1.39)	(2.73)
Farmer	-0.3609**	-0.1925	-0.0295	-0.2697**	-0.1911*	-0.0599	-0.2751**	-0.1679*
	(3.77)	(1.91)	(0.31)	(3.32)	(2.04)	(0.70)	(3.22)	(2.00)
Prosperity	0.1433**	0.1281**	-0.0440**	0.0780**	0.0909**	0.0388**	0.0515**	0.0439**
	(9.98)	(8.98)	(3.45)	(6.61)	(7.33)	(3.27)	(4.15)	(3.60)
Regional unemployment	-1.1288**	-0.6036**	-0.948**	-0.4911**	-1.1050**	-0.4293*	-0.7427**	-0.7252**
	(5.02)	(2.80)	(5.03)	(2.80)	(6.07)	(2.41)	(4.06)	(4.03)
Regional robot density	-0.0825	0.2166	-0.4124**	0.00757	-0.1128	0.04828	-0.2500*	-0.3177*
	(0.62)	(1.41)	(3.30)	(0.06)	(0.93)	(0.42)	(2.00)	(2.56)
Observations	21680	21235	25512	25107	25074	25325	25302	25003
Log likelihood	-17667	-17134	-23049	-52307	-46754	-51034	-45608	-49086
χ^2	2110	2110	949.4	3623	3279	2712	2886	3016

Notes: Regressions estimated by ordered probit, with standard errors corrected for heteroscedasticity. ** and * denote significance at the 1% and 5% level of significance respectively. Country dummy variables included in the regression shown in Table 7.3.

Source: Derived from Eurobarometer surveys 87.1 (2017) (column 1) and 82.4 (2014).

rural areas. There are substantial divides along occupational characteristics. Professionals and managers tend to be more favourable, but manual workers, particularly unskilled manual workers, tend to be less approving. Finally, more prosperous people tend to be more approving. These results tend to be repeated for the other more specific attitudes, which are shown in the final six columns. But there are differences. Approval declines linearly with age for elderly care and both types of AV, whilst the relationship with age for educational robots is an inverted U shape, with approval first increasing then decreasing. The other variables display a greater degree of consistency. For example, people who live in rural areas are significantly less approving for five of the six specific types of robots, the exception being robots in dangerous situations. Professionals and senior managers are more supportive on all six dimensions.

Just two regional variables were significant in these regressions. Firstly, regions with high levels of unemployment tended to be less approving of robots in almost all regressions. Secondly, areas with high robot usage, either in the workplace, home or elsewhere, had more negative attitudes to AVs and to robots used in dangerous situations. Neither regional educational levels nor regional prosperity were significant in any regression. The similarity of the results for robots in different contexts, for example with men and the better-educated consistently more approving could, for example, reflect general attitudes to risk aversion. However, the consistently less approving attitudes amongst manual workers suggests some underlying factor at work, rather than just a calculation of net benefit on each issue. So too does the impact of unemployment, both individual and regional. This common factor could be general disapproval of robots per se, possibly linked to their impact on jobs.

In Table 7.3 we show the country-specific coefficients from these regressions. These are in comparison to Finland, which is included with a zero coefficient throughout. They are ranked in terms of the second equation for general robot approval in 2014. The correlation between robot approval in 2017 and 2014 is quite high at 0.80. There is also a degree of consistency across robot types. Thus Cyprus, Greece and France are particularly hostile across most dimensions, while the Czech Republic and Poland are more favourable. But there are differences between countries across different robot types. Sweden tends to be more positive, apart from towards caring robots, whilst Hungary tends to be less positive, apart from towards AVs.

These differences will be driven by basic attitudes and in part by perceived regulatory differences between countries. But they will also be
driven by the counter-factual, that is, the current state of affairs. For
example, in countries with a high accident rate caused by dangerous
driving, people may be relatively more favourable to AVs. Similarly
with caring robots and the current state of care provision.

7.6 REFLECTIONS

This is the first research that looks at attitudes to robots across a
range of different uses via regression analysis. On balance, people
tend to be slightly in favour of robots, but it does depend upon
their use. This suggests that people are balancing risks and gains,
and that such risks and gains do differ between uses. The differing
nature of the impact of age may well reflect differences in self-
interest between different robot uses. In general, however, the old,
the uneducated, women and those who live outside large towns and
cities – particularly those in rural areas – tend to be opposed on most
dimensions, as are those in manual jobs. In part, these attitudes may
again reflect self-interest; people in rural areas, for example, are less
likely to benefit from robots, and women tend to be more risk-averse
than men. But in part, too, they may reflect ingrained attitudes
perhaps to robots per se, and perhaps to all new technologies. The
significance of the occupational variables, and also regional and
individual unemployment do suggest this to be the case.

The analysis has suggested that attitudes to robots in general are
becoming slightly less supportive. Even amongst the well-educated
there is substantial scepticism about the potential use of robots and
it may be a mistake to think that public scepticism is all down to
ignorance. In addition, the conclusion that familiarity with robots,
as reflected by the regional robot variable, has relatively little impact
on attitudes, suggests ignorance is not that important in impacting
on attitudes. Thus there are genuine concerns about the impact of
robots both in general terms and in their specific uses, which cannot
simply be put down to ignorance and lack of knowledge. Finally,
opinions tend to be polarised with many people either entirely
opposed or entirely in favour. Such a polarisation of views is not
helpful to policy-makers seeking to find a middle ground, as a middle
ground will not satisfy those on either of the two extremes.

Table 7.3   The regression coefficients relating to countries

Country	Approval 2014	Approval 2017	Dangerous situations	Surgical	Education	Care for elderly	AV cars	AV trucks
Poland	0.126	0.369	0.042	0.374	0.389	0.845	0.610	0.525
Estonia	0.171	0.316	0.691	0.336	-0.427	0.091	-0.102	-0.119
Sweden	0.100	0.214	0.746	0.051	0.109	-0.136	0.196	0.202
Bulgaria	0.204	0.208	0.481	-0.008	-0.040	0.201	0.233	0.275
Netherlands	0.301	0.182	0.298	-0.025	0.304	-0.063	0.403	0.288
Lithuania	0.072	0.169	0.447	0.256	-0.654	0.599	0.384	0.293
Czech Rep.	-0.169	0.121	0.152	0.205	0.300	0.528	0.178	0.180
Germany	-0.164	0.098	0.345	0.082	-0.255	0.228	0.004	-0.037
Denmark	0.282	0.097	0.513	0.086	0.069	-0.037	0.288	0.003
Slovakia	0.173	0.073	0.150	-0.390	-0.125	0.039	-0.064	0.044
Latvia	-0.043	0.052	0.532	-0.157	-0.624	0.301	-0.123	-0.004
Finland	0	0	0	0	0	0	0	0
Spain	-0.082	-0.078	0.243	-0.445	-0.276	-0.163	0.001	-0.190
Slovenia	-0.302	-0.082	0.456	-0.450	-0.230	-0.253	-0.029	-0.153
Portugal	0.057	-0.083	-0.022	-0.273	-0.135	-0.242	0.176	0.056
Italy	-0.015	-0.172	-0.017	-0.294	-0.006	-0.306	0.032	-0.171
Belgium	-0.159	-0.182	0.065	-0.173	0.125	-0.008	0.073	-0.118
Austria	-0.360	-0.194	0.086	-0.253	-0.133	0.274	0.178	0.092
Luxembourg	-0.040	-0.204	0.328	-0.515	-0.105	-0.097	-0.049	-0.245

UK	-0.195	-0.206	-0.015	-0.244	-0.311	-0.149	-0.021	-0.224
Croatia	-0.350	-0.208	-0.142	-0.447	-0.308	0.230	-0.067	-0.162
Romania	-0.260	-0.225	0.182	-0.791	-0.445	0.075	-0.037	-0.181
France	-0.280	-0.346	0.021	-0.452	-0.113	-0.056	-0.025	-0.334
Ireland	-0.109	-0.354	0.003	-0.287	-0.304	-0.071	-0.007	-0.209
Malta	-0.219	-0.374	0.039	-0.427	-0.682	0.194	-0.213	-0.297
Hungary	-0.305	-0.432	0.111	-0.318	-0.222	-0.093	0.190	0.147
Greece	-0.376	-0.459	-0.095	-0.824	-0.096	-0.455	-0.170	-0.093
Cyprus	-0.520	-0.563	0.039	-0.907	-0.290	-0.596	-0.437	-0.416

Notes: These are obtained from the regressions reported in Table 7.2. A positive value indicates robot approval and a negative value robot disapproval relative to Finland.

APPENDIX 7.1

Data are derived from Eurobarometer 82.4, November–December 2014 and Eurobarometer 87.1 (2017).

Attitude to robot	*Description of variables and their coding*
Approval	Overall attitudes to robots coded 1 (very negative), 2 (fairly negative), 3 (fairly positive) and 4 (very positive).
Dangerous situations	Response to statement that robots are necessary as they do jobs too hard or dangerous for humans, coded 1 (totally disagree) to 4 (totally agree).

Specific robots	*Coded 1 (totally uncomfortable) to 10 (totally comfortable) to:*
Surgical	Being operated on by a robot.
Education	Using a robot in school for education purposes.
Caring robots	Services and companionship to elderly and infirm people.
AV cars	Travel themselves in a driverless car.
AV trucks	Transport goods in a driverless commercial vehicle or lorry.
Regional robot density	Proportion of people using robots at home, work or elsewhere.

For other variables see Appendix 5.1.

8. Policies to deal with potential problems and to realise the promise

8.1 INTRODUCTION

There are two types of problems. Firstly, the technical ones. How to protect robots against, for example, hacking, and prevent the use of artificial intelligence (AI) to subvert public opinion and undermine democracy? How to ensure that robots do not, in their enthusiasm to carry out their mission, harm human beings and the capital infrastructure? These are difficult, but possibly solvable to varying degrees. Thus, we largely ignore these technical problems that need technical solutions rather than political, economic or social ones, although this does not necessarily mean to imply that technical solutions will be found. These other problems relating to the impact of robots on the economy, society, people and politics will in themselves in many cases be difficult to solve. If, as many claim, robots lead to a host of new jobs we cannot now imagine and there is neither excess unemployment nor large pockets of low-paid workers, then the social problems are limited. If, however, this is not the case, then the social problems are more substantial and their solution probably necessitates fundamental changes to the way society is organised.

Much of this chapter is focused on the latter scenario. But to begin with, we examine the possibilities in the optimistic scenario. The potential benefits of robots should not be underestimated, but for much of the chapter we then focus more on the potential problems. These come under several headings. Firstly, they relate to the economic issues surrounding if not 'the death of capitalism' then at least 'the metamorphosis of capitalism', and the need to funnel spending power to people who are either not working or not working in well-paid jobs. Secondly, there are the social problems related to a declining population and the increasing isolation of people from other people where they live. Thirdly, there are the political problems, albeit with solutions that to an extent depend upon the solutions to the other problems. Fourth, there are the security and privacy problems. Fifth, there are the threats to human beings themselves.

Underlying all of these is the question relating to how far we would want technology to go in terms of new discoveries, particularly, in the context of this analysis, with respect to robots. In a sense it is a redundant question, as to a large extent we cannot control science, with it being an example of the invisible hand at work. Regardless of what the majority of people would wish to happen, what they would wish to be discovered and what would be best for humans, as long as there are individuals, firms or governments working in this area, the boundaries of science will be pushed back in a semi-random manner. We can slow the process down, but it is difficult to stop it. Nonetheless the question is worth asking, if only to understand what in an ideal world would be the state of technology. Would it be a steady-state technology, where science has reached this point and no more? Or would it be along a continually controlled optimal growth path, and if so, what would be the speed of that growth? In understanding this issue, we understand some of the deadweight losses and costs of uncontrolled scientific endeavour.

8.2   MAXIMISING THE POSITIVE BENEFITS

The benefits come under two headings. Firstly, developing a robot-based economy can give a country an economic advantage, possibly in absolute terms, but certainly relative to competitor economies. Secondly, there is the beneficial impact of robots in making our lives better on every possible dimension. These include in hospitals, in caring for the elderly, in dealing with hazardous situations, improving mobility, joining up remote areas and so on. For both headings, the key lies in research in both firms and universities, but increasingly the latter are important. Apart from the general funding of universities, research grant funding can be suitably directed to emphasise robot research. In addition, in order to develop a healthy robotics industry, countries need to invest in the right skills, not just amongst the robot producers, but also the robot users. However, to an extent what is taught is decided by the invisible hand. Individual universities and colleges provide the courses, and students or retraining workers decide which courses to take. The implicit assumptions are that students will choose those courses that maximise their earning and labour market potential and that universities will meet this demand. But it is an imperfect process, subject to lags. Thus government guidance would be useful, not least because we need, in a rapidly evolving world, to be teaching tomorrow's skills not yesterday's. This may call for a thorough revision of what is taught in both schools, colleges and universities.

However, a strong research base is no guarantee that the product will successfully be brought to market. For that we need the entrepreneur. This is the type of fast-growing area that sees the launch of many new firms. We saw repeatedly when looking at the history of robots and the integrated circuit how small start-up firms, based on new research, became major players. But there is a further problem in today's world: the successful new start-up is often swallowed by the large, often for the UK at least foreign, multinational, and in the long run that brings relatively little benefit to the country that developed the research. So, in order to maximise the benefits of robots, countries need to invest in research, provide the right climate where that research can be brought to market and, if possible without becoming protectionist, ensure that the firm does not get taken over by multinationals before it has reached its full potential – if then. This perspective is reflected in work commissioned by the European Parliament, where Nevejans (2016) argues that for the EU, from an economic perspective, robotics should be developed along with increasing aid to businesses, training people with relevant skills, financing universities and so on.

There are other factors to consider. If the fears of the citizen are irrational, then they need reassurance on the benefits and safety of robots and that implies both education and good regulation that people trust. Regulation can thus help and not hinder innovation. In addition, it has to be considered whether letting the market dictate the character of robot innovation is optimal, or whether it will leave gaps in robot development in areas that, although not as profitable to firms as other possibilities, offer greater social returns. This may well be the case, for example with the development of rural robots, and if so, there may be a case for government intervention, if only via grant funding initiatives. In a similar vein, Acemoglu and Restrepo (2018), when discussing AI, have also argued that governments should intervene to control the direction of innovation, focusing more on those aspects of AI that complement labour or create new job opportunities – a similar argument that can also be found in Delvaux (2017).

Currently, some of the most strongly growing areas involve data analytics. This, together with AI and machine learning (ML), has substantial potential for economic growth in, for example, healthcare diagnostics, supply-chain management and sustainable energy management. Realising this potential depends upon access to large data. But unfettered access to large amounts of data, much of it private, can and has been abused by companies. Optimal policy requires striking the right regulatory balance. Also important is access to the right infrastructure; today, that means access to fast Internet download/upload speeds for firms, the public sector and households. In addition, all urban areas, major roads and railways need good fifth-generation mobile (5G) coverage. Much of the future economy will be a digital one and to a large extent the Internet represents the rails on which it will run.

One issue holding back the development and full deployment of robots is their uncertain place within the law. The problems with AI and intelligent robots are particularly severe. What happens when an AI machine makes a decision that results in harm? Who is to blame? Who is legally liable? Who pays the insurance? These unanswered questions are regarded as major obstacles in the way of widespread adoption of AI and intelligent robots. Current law only tends to cover foreseeable damage caused by manufacturing defects. However, with intelligent robots, foreseeing harm is more difficult. This lack of clarity applies to all countries and is in urgent need of resolution. Mady Delvaux, in a 2017 report to the EU parliament, argued that, at this point in time, the responsibility must lie with a human and not a robot, and that there is a degree of responsibility on the trainer or programmer. Further recommendations include setting up an obligatory insurance fund, as with cars. In order to facilitate traceability and hence the allocation of responsibility, and also for other reasons, a system of registration for advanced robots should also be introduced. The more autonomous the robot, the less blame could be attributed to the manufacturer and, in insurance terms, any excess may be covered by the insurance fund. However, her most controversial recommendation in this context lay in suggesting that suitably qualified robots be given the legal status of electronic persons. Progress on these issues is important, and again, in a word we use many times with respect to robots, urgent.

8.3   SOLVING THE ECONOMIC PROBLEMS

The paper by Nevejans (2016) perceives robotics as being an ingredient for economic success and in some cases economic revival. This is probably true in the short term and individual governments have no option but to pursue policies to enhance robotisation. But even in the short term, as we argued in earlier chapters, robot development in one country or region may simply cause unemployment and a lack of prosperity in other regions and countries. Nevejans further argues that the economic advantages offered by robotisation might prompt production, which had been relocated to emerging markets, to return to Europe. That solves or goes some way towards solving Europe's problems, but at the cost of exacerbating the problems elsewhere, particularly in developing countries. But such is the potential for robots to replace jobs that it is not clear that it will even solve Europe's problems. Hence Nevejans also suggests that robots could wipe out several million jobs across Europe, both low-skilled jobs and high-skilled ones such as teaching. Thus it is argued that 'the robotisation and relocation of industry therefore needs to be planned in such a way as to support – rather than supplant – people in the workplace' (ibid., p. 12). This is obvious, but the more difficult question is: how? What does such planning look like? The solution suggested by Nevejans is to finance universities to come up with novel ways of getting round these, and other, problems. Of course, that will help and must be done, but I am sceptical that it will succeed in solving all the problems.

We argued in Chapter 6 that robots have changed the scenario regarding the circular flow of income and that in many cases labour is being replaced in the production function by robots, a form of capital. This gives rise to both a lack of purchasing power in the economy as well as problems with inequality. The question is then how to organise society so that the income of poorly paid workers is augmented to provide them with a minimally acceptable standard of living and also augment the purchasing power of capital in the circular flow of income? One possibility is to tax production and use the tax to augment labour income, that is, a transfer payment. The universal basic income concept is one example of this, which in some versions, as its name suggests, is an income given to all. Other versions could be income dependent (Reed and Lansley, 2016). These redistributive policies will need to be financed primarily by increased taxation on firms, particularly those making large profits directly linked to the introduction of robots. Firms – typically but not just multinationals – are likely to respond by resorting to tax avoidance and even tax evasion, as will richer individuals in society. The political will needs to be in place to deal with these problems and to use AI in fighting against tax avoidance and evasion.

There are already discussions of this in certain countries. Finland, for example, trialled this in 2017, paying a random sample of 2000 unemployed people aged 25 to 58 years 560 € a month, which they received whether or not they found a job. It has the support of some famous names such as Mark Zuckerberg and Elon Musk. Delvaux (2017) too is concerned with the sustainability of the tax and social systems, and also tentatively suggests the possible introduction of a universal basic income. But as a traditional economist I wonder why it has to be paid universally rather than in a manner more targeted to those adversely affected by automation and robots? Another possibility is that each citizen is given an investment credit with which they can acquire shares in their chosen company. Similar reservations apply as to why this has to be to every citizen. On the other hand it could be argued that it might be cheaper to do it this way than checking every claimant's income.

8.4   SOLVING THE POLITICAL PROBLEMS

To an extent when one solves the economic problems, many of the other problems are on their way to being solved. Thus, in a sense there are no purely political solutions to the political problems that will arise. This is important as there are likely to be calls for political reform as the problems mount. This is to misjudge both the problem and the solutions. Having said that, modern technology does pave the way for the individual having more political involvement in a manner that would have been impossible even 50 years ago. Hence political reform may arise from this, but not as a reaction to economic problems.

With the erosion of job markets and the rise in unemployment or lowly paid jobs, social unrest is likely to increase. This increase is likely to be focused within certain age groups and, in many countries, within certain ethnic groups. They are likely to vent their dissatisfaction by voting for extremist parties or new parties, who promise to solve their problems. But given that the problems are being caused by automation, these solutions are unlikely to work. At some point the people will realise the real causes of the problems and at that point they may turn away from the ballot box to direct action, copying the actions of the early Luddites several centuries ago. The immediate reaction of firms will be to protect their machinery and capital with enhanced security, both human-based and increasingly robot- and AI-based. This will lead to increased polarisation within society and increased inequality and still further social unrest. As we have emphasised, there are no political solutions to this; the solutions have to be economic ones.

But there is another problem. Up till now governments have always depended upon people for their income. To a great extent, taxes have been raised in the main part from citizens either directly from their labour income or indirectly from their expenditure. A smaller proportion has also come from firms, which ultimately some of the population own. In the future, non-capital-owning citizens are likely to be net beneficiaries from the state, even more than is the case now. The government will not depend upon them for income; rather, the situation is reversed, and they will depend upon the government for income. Governments instead will ultimately obtain more, and perhaps most, of their income from firms and also the educated elite. Will this change the fundamental system of governance? Will there then be a shift away from satisfying the needs of the ordinary citizen to those of firms and the privileged?

8.5   SOLVING THE SOCIAL PROBLEMS

This is one area where robots may solve more problems than they give rise to. We cannot look at all of these, but we will focus on a number of specific issues. The problems in part arise due to robots substituting for labour in both the household production function and the household leisure function. The first ensures that an individual will be cared for as they grow old and that they are able to continue to live an independent life, without going into care, particularly with the use of exoskeletons. This is very much a positive, particularly in light of an increasingly elderly population where many people have no close relatives to look after them and the state is financially constrained. However, this development does reduce the need to have children and possibly to get married. The second possibility of robots facilitating the enjoyment of leisure also reduces the need for social interaction with other people, at least other people in the immediate spatial neighbourhood, and the need for marriage. In order to counter the negative impact on the number of children, the government payment to people noted above could increase with the number of children up to a certain limit. Unlike child benefit, some payment benefit could be for life, thus increasing the returns children provide to the parents and increasing the incentive to have children and to get married.

There are also concerns about the impact of robots on society. Nevejans (2016) states that once robots become common both in the home and at work, society will change dramatically, even to the extent where people fall in love with robots. Thus, the split between past and future societal models will be fundamental. With robots in the home, there is a risk that they become the norm and people the exception, with the potential of isolating people from other people in all areas of activity, including the individual's personal life. There may also be problems in individuals identifying too closely with robots and because of this there is an argument not to make them too attractive, too lifelike, too lovable and too agreeable. Thus, for example, when maltreated, they might be programmed to react, albeit in a restrained manner. But for the elderly person on their own, the reverse is true. They may have few 'people' other than the robot to depend upon. Hence caring robots need to be targeted at the characteristics of the person being cared for and this suggests that caring robots should be licensed, with the license being granted in a way that is dependent upon it satisfying certain characteristics that vary with the needs of the individual. This potentially applies to all robots that interact with humans and has similarities to the suggestion by Delvaux (2017) that a system of registration of advanced robots should be introduced.

The replacement of humans by robots means that people will not work as much as they do now. But this then raises the question of what people will do. How will people spend their lives? Will it be in permanent rest and relaxation? Will people pursue cultural activities such as music, poetry and drama on one level and gardening, pigeon racing and active sport on another level? Or will they spend all their time watching television – or whatever that has evolved into? The latter may lead to a reduction in human abilities and intelligence and the former to a golden age. This too is a policy area and eventually governments will need policies to encourage active rather than passive leisure activities and leisure facilities that enhance human development, social capital and bonding.

8.6 SOLVING THE MILITARY PROBLEMS

At this point in time, lethal autonomous robots (LARs) may be more likely to make errors, such as killing an enemy combatant who is trying to surrender, than human soldiers. However, robot soldiers are not necessarily bad news for civilian populations and other combatants. Human soldiers often behave badly, either because they *are* bad, or they are affected by the 'fog of war'. The latter should not be relevant for a robot, although there must be concerns about what happens when a fault develops, and the former would depend on their programming. It has also been argued that robot soldiers are not driven by the desire for self-preservation. Robots may also be better at processing new information rapidly and, eventually, with a broad range of sensors, sensing local conditions and identifying potential targets and enemies. This may prevent the inappropriate use of force. The presence of robots on the battlefield, together with human soldiers, may help in the monitoring of the latter's actions, which of itself may lead to a reduction in ethical infractions.

But there is also a downside to this. Being machines, robot soldiers can also be more readily sacrificed in combat situations. A fundamental cost of wars is the loss of life, and in democracies that makes wars unpopular and tends to put a break on their use. Robot soldiers do not come home in body bags and as a consequence the human cost of war, particularly to the country with most robots with the war being fought on foreign soil, will be reduced. This may lead to democracies becoming more willing to engage in wars, before pursuing non-violent alternatives. There is a further problem: wars do not tend to be a solution, they tend to pave the way to a solution. This often means winning the hearts and minds of the people on the losing side. Respect and trust is crucial to this process and the use of LARs may fuel resentment rather than trust. In addition, the assumption that self-preservation will not motivate robots has been called into question by Omohundro (2008), who argues that no matter what its targets are, an intelligent robot would automatically pursue a set of goals that would allow it to achieve those targets. These goals may well include self-preservation.

With respect to LARs, there is sufficient concern, particularly amongst those involved with their development, that it is plausible that action will soon be taken. Thus in August 2017, more than 100 robotics and AI industry leaders signed a letter[1] warning about autonomous weapons and urging the UN Convention on Certain Conventional Weapons to act in order to prevent a high-tech arms race. According to the letter, 'once developed, they will permit armed conflict to be fought at a scale greater than ever, and at timescales faster than humans can comprehend'. The letter goes on to comment that there is not long to act and follows on from previous letters from, for example, Stephen Hawking, Steve Wozniak and Elon Musk along similar lines. But as the letter said, time is indeed short to achieve these objectives.

The question is: how to proceed? There are problems with international arms control treaties. Firstly, they apply to states and transnational terrorist groups, for example, are outside their scope. Secondly, such treaties require state consent, which is not always forthcoming, and states can withdraw from them at any time. For example, the Comprehensive Test Ban Treaty (CTBT) adopted by the UN General Assembly in 1996 prohibits any nuclear weapon test explosion or any other nuclear explosion. But it still has not come into force as it has not been ratified by nine key countries, including the USA. Binding prohibitions or restrictions on all member states can be imposed by United Nations Security Council (UNSC) resolutions, such as UNSCR 1540. This imposes binding obligations on all states to adopt legislation to prevent the proliferation of nuclear, chemical and biological weapons, and their means of delivery,[2] although as we have seen in recent years with respect to Syria and North Korea this has not been wholly successful. Despite the need for some treaty, there is also pessimism about the likelihood of achieving one, due in part to the difficulty in negotiating it. Thus Marchant et al. (2015) observe that at least in the short term, formal hard law treaties may not be practicable to achieve. But the attempt has to be made.

---

[1]　https://www.theguardian.com/technology/2017/aug/20/elon-musk-killer-robots-experts-outright-ban-lethal-autonomous-weapons-war.

[2]　http://www.un.org/en/sc/1540/about-1540-committee/general-information.shtml.

8.7 A POLICY FRAMEWORK FOR NATIONAL GOVERNMENTS

Despite the threats robots pose to jobs, a national government has little choice but to pursue robot development. If they do not, other countries will and their economies will suffer. Hence governments should pursue policies to enhance the positive impact of robots on their economies, as argued earlier. The solution to many of the problems that will then arise because of robots and AI lies in strengthening the legal framework surrounding robots and AI and laying down principles of privacy, as well as ethical considerations. Some of this is best done at the international level, but national governments also have a role. Delvaux (2017) makes a series of recommendation for the EU to adopt. Whilst not a national government as such, this does provide a blueprint for national governments to build upon. The first point made, which is surely right, is that policies need to be implemented now. Collingridge (1982) observed that not enough is known about a new technology, prior to its development, to guide its governance, but that once we have that knowledge through its deployment, it is often too late to regulate it. By this time the commercial exploitation of the technology is too far down the line to reverse. This is not a mistake we can afford to make with robotics; we need to regulate now, and then frequently adapt those regulations as further knowledge becomes available.

Many of the Delvaux (2017) recommendations in the document are generic and have relevance across several policy concerns. Others are more specific in their impact. Amongst the former are:

- a guiding ethical framework for the design, production and use of robots
- a charter consisting of a code of conduct for robotics engineers, and for research ethics committees when reviewing robotics protocols and of model licences for designers and users
- the need to test robots in real-life scenarios is essential in order to identify any risks. The document also emphasises that the testing of robots in cities and on roads raises many problems and requires effective monitoring.

In order to help with this, the establishment of a properly staffed and financed European agency for robotics and artificial intelligence is recommended. This will provide the technical, ethical and regulatory expertise needed to support the policy-makers. It will enable countries to both maximise the benefits of the new opportunities and meet the challenges arising from robots. It will also help to protect society from the potential harm robots may cause. We would argue that such agencies should not just promote the use of robots, research into robotics and their regulation, they should also be responsible for foreseeing the social and economic consequences of robots and AI. A single agency would allow for economies of scale and the informing of ideas and policies in all parts of the agency. Hence new research would be informed by ethical considerations and vice versa. The risk is that one part of the agency would become dominant, to the detriment of the other parts. On balance it should be possible to establish a governance structure to ensure a single agency fully meets all the elements in its remit.

The agency also needs to be people-focused and not just robot-focused. If robot carers provide risks to individuals in terms of social isolation, the solution is not necessarily just to tweak a robot or to ban it in certain usages, it is also to train the human carers to compensate for that and to make the caring agencies aware of the problem. This is true not just for caring robots, but also for many other robots, especially those in transport, policing and education. The advice to the education agency, for example, needs to be of two types, firstly focusing on likely skill needs and shortages in the near future, and secondly on the impact of robots on the delivery of education itself, with the impact on both the student and the school or college. In this, the robot agency would benefit from liaising with university departments. We have seen that universities have played a critical part in robot development across the world. But the emphasis has been much more on technical and scientific aspects, rather than tapping into social science skills. This is reflected in the far greater number of university robot centres in science and engineering than in the social sciences.

8.8　AN INTERNATIONAL DIMENSION

Some of the problems are best tackled at the international level. This is typically true of military robots, as already discussed. It is naïve to expect a government not to use a specific type of robot soldier or weapon if potential enemies have access to these. But international regulation could also help in controlling future research directions. Once a new discovery is made, it is difficult to stop its use spreading. One example is the UN Convention on Certain Conventional Weapons, referred to earlier. Another, in a different, non-military, context is the Convention on Biological Diversity (CBD), which has been signed or ratified by almost 200 countries. It is an international legally binding treaty and has three primary goals: (1) the conservation of biodiversity; (2) the sustainable use of biodiversity; and (3) the fair and equitable sharing of the benefits arising from the use of genetic resources. This is the kind of treaty that might, for example, be relevant for cybernetics.

However, as discussed earlier, there are limits to how far such international cooperation can go and, despite the large number of signatories, the USA is not a signatory to the CBD. Thus, again, it must be doubted whether international cooperation can prevent all harmful robotics development. But, in the absence of any better alternative, the attempt must be made and it must be recognised that it can be partially effective. The alternative to hard law is soft law and it is possible that standards bodies such as the International Electrotechnical Commission (IEC) and the International Organization for Standardization (ISO) could expand the number and scope of standards related to robotics. Some of their existing standards are technical, relating to interoperability and tests, some are taxonomic, but almost one-third are focused on safety. However, such clear-cut distinctions are a little misleading and any standard may facilitate safety as well as efficiency and economies of scale. Clearly in this, they go some way to dealing with people's concerns, particularly regarding safety aspects. But they still, inevitably, leave a lot of gaps.

There is also a problem with take-up. For example, ISO 13482 is focused on personal care robots and lists typical safety-related functions, such as speed and collision avoidance. However, Guiochet et al. (2017) comment that relatively few robots have been ISO-certified and at this time a safety culture is lacking. There are other standards. The British Standards Institute's BS 8611 is targeted at manufacturers and designers when designing robots. It provides: (1) guidelines that help identify potential ethical harm; (2) guidelines on safe design, protective measures and information for the design and application of robots; and (3) builds on existing safety requirements for different types of robots, including industrial, personal care and medical robots. The ethical issues and hazards include the dehumanisation of humans and over-dependence on robots.

8.9   A MATTER OF URGENCY

We have seen that innovation in part is influenced by government policy, by laying down the rules and by providing funding for specific areas of research. But we have also seen that innovation follows its own course, driven by the invisible hands of scientific curiosity and the pursuit of profits. Thus, even if a government wanted to fully control the pace and direction of robotics research it would face great difficulties and it would probably be impossible. This is particularly the case when there is not just one government but many, and many in countries with sufficient resources to pursue this research. If a country chooses not to do this, it risks becoming impoverished relative to others. Hence a government might want to put a limit on the development of job-replacing robots, but that would be difficult. Firms pursuing profits will develop and market job-replacing and other robots and they will be used, jobs will be replaced and they will change society. Hence, the expansion of robotic influence is almost inevitable.

But that does not mean governments are totally powerless. We can to a degree control some aspects of the impact of robots by the form of regulation discussed earlier and also by international treaties. Delaying regulation of robots until we know all or most of the impact is too late. By this time the commercial exploitation of the technology is too far down the line. This is not a mistake we can afford to make with robotics; we need to regulate early and then adapt those regulations as further knowledge becomes available. This is true both for robot regulation and controlling robot research and development. Now, at the beginning of this revolution, is the right time to ask what we can do and what we should do to minimise the adverse impact of robots and to maximise their beneficial impact. One possibility lies with robot programming. Should Asimov's laws, or some other laws of robot behaviour, be programmed into all robots, with failure to do so by the developer and the user being subject to criminal sanctions? Or should these laws simply be focused on the manufacturers? Even then, robots driven by AI may decide to autonomously revise their programming.

8.10   THE OPTIMAL LEVEL OF ROBOTS

Ideally we would want robot research and development to be targeted in a manner that maximises the welfare of human beings, now and in the future. There are two questions to consider in this context. Firstly, what is the optimal level of robotisation and, secondly, how quickly should we move to this position? It is, as always, a mixture of trading costs and benefits and it is not a simple answer to apply to the whole sphere of robot development. For each area of robotics, if the gains outweigh the costs then robot development should be pursued. A traffic-light system – green for go, amber for caution and red for stop – may be employed. Thus, robots in areas such as caring could well be given a green light, albeit a qualified green light, with adverse effects minimised by appropriate design, regulation and licensing. In terms of the speed of development, the slower development proceeds and the more time carers, the cared for and the regulators have to adjust, the lower will be the costs, but the slower will be the benefits to appear. AVs and drones may also be given qualified green lights. Areas where a red light might be applicable may include the military and surveillance, provided this applies to all governments. The latter threatens the privacy of us all. On the other hand, robot sentries that could guard the workplace, the home and the individual from potential harmful humans and robots, including surveillance robots, should be given a green light. Developments that fundamentally change human beings, such as exoskeletons and brain computer interaction, may also be subject to a red light, but it is context-specific, and when they facilitate mobility amongst those who currently have difficulty in moving or speaking they could be given an amber light. Such exoskeletons should be focused on increasing mobility, but not strength nor speed. Similarly exoskeletons in the workplace, in most cases, should be encouraged. Another area where the lights could well be red is cybernetics.

In terms of robots taking jobs, an amber light might be appropriate. There is a case for stopping some robot development now, or at the very least slowing it down substantially. Yes, robots open the possibility for greater prosperity and, provided we can solve the redistribution problem, this could improve the living standards of everyone. But the literature on wellbeing shows that happiness is not just linked to income. We know the unemployed tend to be less happy than others, even given their income level. Simply increasing people's income, but taking away their work and their social interaction with work colleagues may well not make people happier in the long run, but the reverse. Nor will replacing skilled jobs, in the wider sense of the word, with more mundane jobs where the individual has less chance to express their individuality, increase wellbeing. The above is subjective, but if it were possible to put barriers in the path of some aspects of robot development, we might well wish to do so. This is consistent with Joy's (2000) argument that research into the development of dangerous technologies should be limited. Having said that, it may well not be possible, as already indicated, to stop such research and development.

9. A changing world of innovation

9.1 INTRODUCTION

We began this book by looking at innovation. We conclude it with some discussion on what light the development of robotics sheds on this process of innovation. The first point to note is that Schumpeter's observation that new discoveries may lie dormant for several years until picked up by hard-pressed firms searching for profits is now less true. Many of today's firms are innovation-focused. They are actively searching for new discoveries, have their own R&D departments and are in close collaboration with universities. Reflecting this, Sachs (2018) shows that R&D's share of GDP has increased from around 1.3 per cent to 2.6 per cent, from the early 1950s to today. Simultaneously with this, the stock of intellectual property rights (IPR) has risen from around 4.5 per cent to 14 per cent of GDP. Sachs argues that this shows that the economy, and thus many firms, have become far more science-intensive and innovation-focused. The second point is that our analysis has revealed several factors that, if included in the mainstream models, are not emphasised enough. Most importantly they tend to be partial models focusing on the process of innovation, not the subsequent impact on society and the economy and then the feedback from such an impact back onto innovation. The traditional view is, in the main, that innovation is a good thing, to be encouraged without too much consideration of its impact on society. That is not satisfactory. Research changes society, changes people and changes the economy. There may be a gap in the market that a new innovation fills, but in changing society, new gaps and new opportunities appear as well as new threats. Our models must capture this wider impact, and in particular the impact of innovation on wellbeing. To ignore that is to ignore the whole point of innovation. Apart from this, there are several other factors that we seek to emphasise.

Firstly, innovation is increasingly the combination of different sciences. To an extent this has always been the case. The railways required innovations on several different fronts. But modern-day robots, for example, especially once we advance beyond industrial robots, are the children of several diverse scientific disciplines, particularly specialised robots such as surgical robots or nanorobots. Secondly, the role of the university has changed. It has become part of big science, and in some respects university research centres and university academics are part of the research divisions of firms. Thirdly, the role of government is not just to provide the institutional framework that underpins both research and development and to facilitate the interaction between universities and firms; governments are financing much of the research either directly, as with the military, or indirectly through research grants. Fourth, traditional models do not pay sufficient attention to public attitudes. Lack of public approval for a new technology can delay, perhaps indefinitely, the speed with which the innovation is diffused through the economy. In this chapter we examine each of these developments in turn.

9.2   MULTI-DIMENSIONAL INNOVATION

There has always been an interaction between different academic fields, for example engineering and science, where science enables new engineering breakthroughs that then facilitate further scientific breakthroughs, as with the microscope. However, this interaction is becoming more common and more planned. Increasingly innovation is multi-dimensional, that is, it combines researchers from several disciplines. With robots, as we have argued, this is obviously the case. Outside of robotics there are other examples. The science of nanotechnology, for example, is the result of interaction between the more traditional fields of chemistry, physics, mathematics, biology and engineering sciences (Hacklin et al., 2009). This combining of different science disciplines means there needs to be a coordinating mechanism, and perhaps coordinating or integrative research, to combine the different technologies together.

Bainbridge and Roco (2016a) suggest several reasons why such collaboration may be increasing. Firstly, as we become wealthier we can afford bigger and more complex research projects, with increasing numbers of scientists in an increasingly complex division of labour. Secondly, the opportunities for connections between two scientific fields increase exponentially with the number of fields, and the number of fields has been growing in recent decades with the appearance of new disciplines such as information technology (IT). On the policy side, this is leading governments to develop strategies to promote inter-disciplinary and inter-industry collaborations. For example, in South Korea this is being done with a specific focus on medical health, safety, energy and environment (Pak and Rhee, 2016).

One version of this coming together of different disciplines to pursue common research agendas is called convergence research. It is argued that in doing this, scientists' ways of working become increasingly intermingled and integrated, hence overcoming the integration problem referred to earlier. This can even lead to new frameworks or disciplines. The convergence paradigm intentionally brings together intellectually diverse researchers to develop effective ways of working and communicating. It has been suggested that this should not just be limited to engineering and the sciences, but that convergence must integrate ethical analysis and the social sciences to focus on both the intended and potential unintended consequences of the research (Bainbridge and Roco, 2016b). These social science disciplines can also help in optimally implementing the innovation to maximise the beneficial impact on society and the economy.

But there is another aspect to this process. Not only are different technologies being combined in a single innovation, innovations commercialised in one industry or market are directly impacting on other industries and markets. There have always been knock-on effects of major innovations on other industries, as with the railways, but we are now looking at when one industry's products or services are actually sold or used in another industry's, thus eroding clear market boundaries (Lei, 2000). Apart from changing markets, such convergence can also lead to mergers between firms in the related industries. There is also the possibility of new innovation at the interface of the two markets that combines technologies from both. In a sense this is the ex-post collaboration of science disciplines, but probably not done so much within universities as within firms. Technological convergence is where different technologies evolve towards performing similar tasks and competing in similar markets. An example lies with digital convergence. This allows the same content to be accessed on different platforms (TV, smartphone, PC and so on). There is also a functional overlap in the use of different networks and terminals. For example, one can access the Internet on the mobile phone, the PC or the TV, and can converse with others using either a mobile phone or a PC.

9.3 THE CHANGING ROLE OF UNIVERSITIES

What is the role of the universities in all of this? It is apparent that in the past many innovations were done by individuals working alone. In the case of Arkwright it was a working man who saw the potential for advantage. James Watt, who was friends with the economist Adam Smith, was initially based at Glasgow University, although as an instrument-maker, not as an academic, when he worked on the steam engine. But often the scientific advances were made by the gentleman scientist, a person, usually a man in those times, who had the means to pursue their scientific curiosity. Sir George Cayley is one example, who at the beginning of the nineteenth century was a pioneer of aeronautical engineering. There were of course some examples of major breakthroughs made by university academics, particularly as we move to more recent times, for example the work of Marie Curie and Ernest Rutherford. But this has become more common in recent years, although the lone inventor has far from disappeared.

Throughout this book we have highlighted the research being done in universities, frequently, particularly today, in dedicated labs. Specifically, with respect to robots, the role of universities has been fundamental to their development since the early 1950s. The universities are often working closely with firms on this, either in the form of individual scientists or by the firm sponsoring the research. In many respects universities have become the research arm of large, particularly multinational, firms. Robot research involves bringing together knowledge from many different technologies. Universities, almost uniquely, have access to this diversity of expertise in-house, which may be one reason why firms have chosen to become so closely linked with universities. For example, in the UK, King's College Centre for Robotics Research is funded by the UK Research Councils, the EU and companies such as Ford, RU Robotics, Shadow Robotics, Unilever Research Ltd, Yasakawa Electric Co (Japan), Boal BV (Holland) and Tele-Spec Ltd. In addition, many of the firms begin as university, or individual university scientist, start-up firms, as with Victor Scheinman who started Vicarm Inc. to produce his robot arms. This development is not limited to robotics. In the pharmaceutical industry in 2012, Bristol-Myers Squibb announced collaborations with ten cancer research institutions in Europe and the USA. GSK has had several partnership deals including ones with University College London (UCL), Cambridge and Nottingham Universities in the UK, as well as Yale in the USA (Hudson and Khazragui, 2013).

This raises serious questions of academic freedom and independence, and also whether the universities are getting a fair return on their expertise, particularly when the firms involved are foreign ones and the research will not benefit the country where the university is based in terms of creating employment. This involvement with firms is also blurring the lines between the roles of firms and academic researchers in innovation. Firms have always been involved with research and, as with the integrated circuit, have made some critically important breakthroughs. But increasing reliance on universities as a research resource is also changing firms, who had previously preferred to keep research in-house. In part this change is being driven by cost considerations. It is often cheaper to finance research in publicly subsidised research institutions than for firms to have their own research facilities. Apart from funding from firms, university research is often funded by government, in many cases indirectly through different forms of research council grant, as in the case of King's College Centre for Robotics Research. Governments, too, often set the research agenda for universities.

Universities also play a key role in the secondary research. This is not research directly related to the eventual production of the innovation itself, but research that helps assess its use, safety and impact on people and society. This secondary research often highlights potential or actual problems or opportunities, and is then a spur to governments to introduce regulations and regulatory bodies to counter the problems or exploit the opportunities. This secondary research is often not done by academics of the same disciplines who were involved in the primary research activity but may be, and frequently are, social scientists.

As a consequence of all these changes, universities too have changed. In the twenty-first century universities are seen as a key part of a country's research capacity that can be used to secure the economic prosperity of the country. In research terms, the emphasis has switched from blue skies research to more targeted research,

solving specific problems or in specific areas that government thinks
are important and that facilitates firms to be internationally competitive. In some countries this is affecting the entirety of academic disciplines, not just the sciences and engineering that have the most direct
impact on innovation. It is debatable whether this is wholly a good
thing. The pressure is on academics to publish and publish regularly
and for that published research to have societal or economic impact.
Hence the days when an academic could spend 20 years researching
a single issue with few publications may well be gone.

Finally, modern universities have mainly organised their structure
along the 'tree of knowledge'-type model (Lenoir, 1997), according
to which, knowledge is split into branches (for example, sciences,
social sciences), then into major disciplines and thence into sub-
disciplines and specialities. The growing inter-disciplinary nature of
science is challenging this organisational structure to meet the needs
of the day. One answer to this challenge is research groups or centres
that combine the different sciences. These tend to be technology-
focused rather than discipline-focused. But universities need to give
them greater breadth and make them more inclusive than is currently
often the case, including the social sciences as well as the sciences,
with the former being partially focused on the intended and potential unintended consequences of the research. Such technology-
focused inter-disciplinary centres might engage in teaching as well
as research. This is beginning to happen. A number of universities,
for example the University of California, Los Angeles, and Seoul
National University, are creating multi-disciplinary campus spaces
(Pak and Rhee, 2016).

9.4   THE CHANGING ROLE OF FIRMS INVOLVED IN INNOVATION

Innovation, particularly of the type we have studied in this book, is
changing firms in the way they combine capital and labour, changing
the concept of labour, changing the way firms market themselves
and changing the conditions under which they can hire labour. For
example, the growth of zero-hour contracts would have been more
difficult without the phone, particularly the mobile phone, enabling
the firm to instantly contact workers (Speak, 2000). Innovation is
also leading to new firms in new industries, although that has always

been a consequence of innovation. But it is changing the nature of innovative firms. Typically they would do research in a specific scientific discipline in-house or in combination with a university or research centre. This is still the case for the majority of innovative firms, but there are some – particularly multinationals – that are engaged in integrative or convergent innovation and are set up so as to organise this. Their task is in part to integrate the different strands of scientific research and technologies into a new form of technology. Their task, too, is to be aware of potential areas for them to move into. In addition, just as researchers build upon the work of other researchers, so firms build upon the work of other firms – albeit somewhat hindered by the patent laws. But in many cases, firms today are combining with each other at the very beginning of the research process, in part because of the increasingly complex nature of innovation, requiring expertise and skills from different disciplines and industries. Finally, as mentioned previously, robots and AI are replacing people in the factory and the office. This is increasing the capital–labour ratio and again fundamentally changing the nature of the firm. The firm used to combine labour and capital in the production process. In many firms the role of labour is declining and in some cases declining rapidly.

9.5   THE CHANGING ROLE OF GOVERNMENTS IN INNOVATION

Traditionally the role of governments has been the background one of ensuring that the proper infrastructure, legal framework and regulations are in place. Thus, with the railways, governments enabled acts of Parliament to be passed, facilitating railway growth. More recently, recognising the need for trained technical people, they have encouraged the growth of universities and university courses in specific disciplines, and similarly for schools. Regulations protect the public from dishonest firms and poor-quality products, but in doing so they also create a market for good-quality products and honest firms. The free market is good at producing innovation. It is not so good at minimising its adverse impacts. That is the role of governments, although, with respect to both compensating the losers and regulation, it is frequently something they do late in the day.

The existence of knowledge spillovers results in incomplete

appropriability of R&D. As a consequence, the equilibrium level of R&D will be lower than the socially optimum level (Spence, 1984). Strengthening IPR through, for example, the legal framework can correct for this, but at the cost of duplicating R&D activities in the form of a patent race. This then provides a rationale for more direct government intervention. In line with this, the role of governments is becoming increasingly proactive in supporting R&D. Thus, they work to bring different agents together, such as firms and researchers, or different firms. Science parks or technopoles are a particularly high-profile example of this, and may include within their boundary all the agents who facilitate innovation (Felker, 2009). This also provides a rationale for governments setting research priorities. Furthermore, there is the possibility that government R&D projects may benefit private R&D. David et al. (2000) suggest this may happen because of: (1) learning effects that enhance the ability of private firms to obtain the latest scientific and technological knowledge; (2) subsidising the use by the private sector of research facilities; and (3) signalling future public sector demand to the private sector.

9.6   THE ROLE OF PUBLIC ATTITUDES

Public attitudes impact on the extent to which a product/service is taken up and the speed with which it is taken up. If a large section of the public is hostile, the innovation may be stalled. This has been the case with genetically modified (GM) foods, where hostility, certainly outside the USA, has severely limited their development. For this reason, governments often try to mould public opinion in favour of a technological development they feel to be important. But the best way to do this may be to have a good regulatory framework. Innovation involves uncertainty and often potential risk. Strong regulatory institutions can give confidence to the citizen and consumer in the innovation. However, public attitudes are not static. They themselves will be influenced by the innovation as some of them acquire it, knowledge will increase and, for better or worse – often the former – attitudes to the innovation will change. But attitudes are also influenced by the country's basic culture and social values and the innovation may also change those. For example, the Internet and social media have opened up new sources of information for individuals. People are also influenced by fiction.

9.7   AN EXTENDED MODEL OF INNOVATION

The case studies in Chapter 1, as well as the evolution of robots that we looked at in subsequent chapters, reveal a much more complex model of innovation than the models of innovation we looked at in that first chapter. Every model of innovation is a simplification of what actually happens, indeed every economic model is a simplification of the real world. But it is clear that these models miss out, or fail to emphasise sufficiently, several important aspects. We illustrate a more complex model in Figure 9.1. The first point to note is that there was no single critical area of basic science that led to robots. Robots, particularly once we advance beyond industrial robots, are the children of several scientific disciplines. The science of these disciplines needs at some point to be combined. In the standard model we have translational research, which is research that translates the basic research into a form suitable for a new product. In our model we include this as part of the design and engineering phase. Integrating research is more than translational research, it is concerned with combining research from different disciplines. The difficulties can be substantial, not least because different academic disciplines have different cultures, although with convergence that may change.

Secondly, it is apparent that the models of innovation are only partial. They are embedded within the economy and society and yet they largely ignore or take as given the values and structure of the economy. However, innovation interacts with both of these in a recursive loop. A new innovation meets the needs of society, or at least part of society, but it also changes society, and that in turn opens up new possibilities for innovation. Public attitudes are critical in impacting on governments and governmental regulation and in determining the size of sales of any product that comes to market. But governments do not act in isolation from the rest of the world. International standards – both specifically as with ISO and IEC standards, and more generally as with denoting the position and views of other governments and other people across the world – impact on what governments can do, often by treaty obligation. Such standards also impact on public perceptions and values. If a company has an internationally recognised standard, then it is more likely to sell to the public, sell to firms, and, critically, to have a greater presence in export markets.

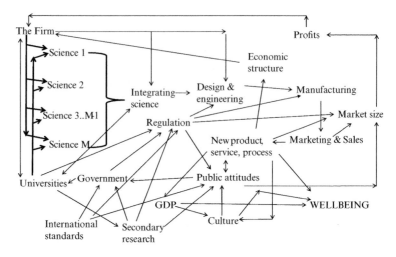

Figure 9.1   The almost complete helix

Governments not only impact on innovation through regulation. They also set the research agenda for universities, part-finance universities and their research, and directly engage in research, particularly research linked to the military, although this is not shown in Figure 9.1. In doing this they are cognisant of the preferences of their citizens, particularly in a democracy, and what they perceive as in the best interests of the country and its economy.

The role of research, particularly university and regulatory agency research, extends beyond the first stages of innovation to a secondary stage. This is concerned with the impacts, both direct and indirect, of the innovation and technology on society, the economy, the environment and on people. The word secondary suggests that it chronologically follows the first research. In a sense this is inevitable. Scientific curiosity is sparked by real-world events and real-world developments. It is when the social and economic impact of the technology begins to be visible that much secondary research is begun. This is perhaps inevitable, as we have said, but it is also regrettable, as the time to take regulatory action is before the impact becomes visible and before the social problems start to arise, as it is much harder to control problems once people have changed their behaviour than before they change it. This entails advance foresight and speculation, and often this appears sensationalist at the time.

Nonetheless this is still an important function of universities and the regulatory agencies.

There are M scientific activities involved with this research and these need to be integrated together before we have research that can form the basis for innovation. The integrating science is done by the firm and in part by the universities, but this is also impacting on the structure of universities. M could of course be equal to one – and often is – in which case there is no integrative science and the model is much simpler. The basic science can be done in universities or the firm or a combination of both. Governments too can directly engage in research. The core of the model is then similar to other models of innovation revolving around profits, design and engineering, manufacturing, marketing, etc. But it also includes the impact on wellbeing as well as GDP. The aim of government policy should not be to create an innovative economy, but to encourage innovation that benefits society and the economy. Secondary research, which in Figure 9.1 is done in universities, is targeted on the wider impact of the innovation. Once secondary research has highlighted a problem, governments react and the regulators react, but so does the culture, as for example with respect to attitudes to smoking once the health effects became highlighted. This culture impacts on public attitudes, as do standards. Public attitudes have a two-way link with new products: a new product can change attitudes, but public attitudes can also modify the product. Finally, innovation impacts on the firm, in the model via the economic structure. We name Figure 9.1 'the almost complete helix' because it includes more than traditional models, but inevitably still misses things out. We do not include, for example, the role of human capital and the role of universities in generating human capital. This is important both in terms of providing a skilled labour force to work in the firms and also in impacting on public attitudes and public abilities. The model also does not directly include the role of fiction; that is included in 'culture'.

References

Acemoglu, D. and D. Autor (2011), 'Skills, tasks and technologies: Implications for employment and earnings', in O. Ashenfelter and D. Card (eds), *Handbook of Labor Economics*, Vol. 4B, Amsterdam: North Holland, pp. 1043–1171.

Acemoglu, D. and P. Restrepo (2017), 'Robots and jobs: Evidence from US labor markets', NBER Working Paper No 23285.

Acemoglu, D. and P. Restrepo (2018), 'Artificial intelligence, automation and work', Paper presented at the INET/IMF Conference on the Economics of Artificial Intelligence in Washington, DC.

Acemoglu, D. and J. Robinson (2012), *Why Nations Fail: The Origins of Power, Prosperity, and Poverty*, New York, NY: Crown Publisher.

Advincula, A.P., X. Xu, S. Goudeau and S.B. Ransom (2007), 'Robot-assisted laparoscopic myomectomy versus abdominal myomectomy: A comparison of short-term surgical outcomes and immediate costs', *Journal of Minimally Invasive Gynecology*, **14** (6), 698–705.

Alemzadeh, H., J. Raman, N. Leveson, Z. Kalbarczyk and R.K. Iyer (2016), 'Adverse events in robotic surgery: A retrospective study of 14 years of FDA data', *PloS one*, **11** (4), e0151470.

Allen, R.C. (2009), 'The industrial revolution in miniature: The spinning jenny in Britain, France, and India', *The Journal of Economic History*, **69** (4), 901–927.

Arntz, M., T. Gregory and U. Zierahn (2016), 'The risk of automation for jobs in OECD countries: A comparative analysis', OECD Social, Employment, and Migration Working Papers No 189.

Asimov, I. (1950), *I, Robot*, Garden City, NY: Doubleday.

Atack, J., F. Bateman and R.A. Margo (2008), 'Steam power, establishment size, and labor productivity growth in nineteenth century American manufacturing', *Explorations in Economic History*, **45** (2), 185–198.

Autor, D.H. (2015), 'Why are there still so many jobs? The history and future of workplace automation', *Journal of Economic Perspectives*, **29** (3), 3–30.

Autor, D.H. and D. Dorn (2013), 'The growth of low-skill service jobs and the polarization of the US labor market', *American Economic Review*, **103** (5), 1553–1597.

Autor, D.H., F. Levy and R.J. Murnane (2003), 'The skill content of recent technological change: An empirical exploration', *Quarterly Journal of Economics*, **118** (4), 1279–1333.

Autor, D.H., D. Dorn, L.F. Katz, C. Patterson and J. Van Reenen (2017), 'The fall of the labor share and the rise of superstar firms', NBER Working Paper No 23396.

Bainbridge, W.S. and M.C. Roco (2016a), *Handbook of Science and Technology Convergence*, Cham, Switzerland: Springer International Publishing.

Bainbridge, W.S. and M.C. Roco (2016b), 'The era of convergence', in W.S. Bainbridge and M.C. Roco (eds), *Handbook of Science and Technology Convergence*, Cham, Switzerland: Springer International Publishing, pp. 1–14.

Bansal, P., K.M. Kockelman and A. Singh (2016), 'Assessing public opinions of and interest in new vehicle technologies: An Austin perspective', *Transportation Research Part C*, **67**, 1–14.

Barbash, G.I. and S.A. Glied (2010), 'New technology and health care costs: The case of robot-assisted surgery', *New England Journal of Medicine*, **363** (8), 701–704.

Bedaf, S., G.J. Gelderblom and L. DeWitte (2015), 'Overview and categorization of robots supporting independent living of elderly people: What activities do they support and how far have they developed', *Assistive Technology*, **27** (2), 88–100.

Benzell, S.G., L.J. Kotlikoff, G. LaGarda and J.D. Sachs (2015), 'Robots are us: Some economics of human replacement', NBER Working Paper No 20941.

Bogue, R. (2016), 'The role of robots in the battlefields of the future', *Industrial Robot: An International Journal*, **43** (4), 354–359.

Bornhorst, F. and S. Commander (2006), 'Regional unemployment and its persistence in transition countries', *Economics of Transition*, **14** (2), 269–288.

Boys, J.A., E.T. Alicuben, M.J. DeMeester, S.G. Worrell, D.S. Oh, J.A. Hagen and S.R. DeMeester (2016), 'Public perceptions on robotic surgery, hospitals with robots, and surgeons that use them', *Surgical Endoscopy*, **30** (4), 1310–1316.

Bresnahan, T.F. and M. Trajtenberg (1995), 'General purpose technologies "Engines of growth"?', *Journal of Econometrics*, **65** (1), 83–108.

Bringsjord, S., J. Licato, N.S. Govindarajulu, R. Ghosh and A. Sen (2015), 'Real robots that pass human tests of self-consciousness' in *Robot and Human Interactive Communication (RO-MAN), 2015 24th IEEE International Symposium*, Piscataway, NJ: IEEE, pp. 498–504.

Broadbent E., R. Stafford and B. MacDonald (2009), 'Acceptance of healthcare robots for the older population: Review and future directions', *International Journal of Social Robotics*, **1** (4), 319–330.

Brynjolfsson, E. and A. McAfee (2014), *The Second Machine Age: Work, Progress, and Prosperity in a Time of Brilliant Technologies*, New York, NY: W.W. Norton.

Brynjolfsson, E. with T. Mitchell and D. Rock (2018), 'What can machines learn, and what does it mean for occupations and the economy?', Paper presented at the INET/IMF Conference on the Economics of Artificial Intelligence, Washington, DC.

Causo, A., G.T. Vo, I.M. Chen and S.H. Yeo (2016), 'Design of robots used as education companion and tutor', in S. Zeghloul, M. Larib and J.P. Gazeau (eds), *Robotics and Mechatronics: Mechanisms and Machine Science, vol 37*, Cham, Switzerland: Springer, pp. 75–84.

Clynes, M.E. and N.S. Kline (1960), 'Cyborgs and space', *Astronautics*, **5** (9), 26–27, 74–76.

Coeckelbergh, M. (2010), 'Robot rights? Towards a social–relational justification of moral consideration', *Ethics and Information Technology*, **12** (3), 209–221.

Coeckelbergh M., C. Pop, R. Simut, A. Peca, S. Pintea, D. David and B. Vanderborght (2016), 'A survey of expectations about the role of robots in robot-assisted therapy for children with ASD: Ethical acceptability, trust, sociability, appearance, and attachment', *Science and Engineering Ethics*, **22** (1), 47–65.

Coffé, H. and B. Geys (2007), 'Toward an empirical characterization of bridging and bonding social capital', *Nonprofit and Voluntary Sector Quarterly*, **36** (1), 121–139.

Collingridge, D. (1982), *The Social Control of Technology*, New York, NY: St. Martin's Press.

Cundy, T.P., H.J. Marcus, A. Hughes-Hallet, A.S. Najmaldin, G.-Z. Yang and A. Darzi (2014), 'International attitudes of early adopters to current and future robotic technologies in pediatric surgery', *Journal of Pediatric Surgery*, **49** (10), 1522–1526.

D'Andrea, R. (2012), 'Guest editorial: A revolution in the warehouse – A retrospective on kiva systems and the grand challenges ahead', *IEEE Transactions on Automation Science and Engineering*, **9** (4), 638–639.

Dauth, W., S. Findeisen, J. Suedekum and N. Woessner (2017), 'German robots: The impact of industrial robots on workers', CEPR Discussion Paper No 12306.

David, B. (2017), 'Computer technology and probable job destructions in Japan: An evaluation', *Journal of the Japanese and International Economies*, **43**, 77–87.

David, P.A., B.H. Hall and A.A. Toole (2000), 'Is public R&D a complement or substitute for private R&D? A review of the econometric evidence', *Research Policy*, **29** (4), 497–529.

Delvaux, M. (2017), 'Report with recommendations to the Commission on Civil Law rules on robotics', Report for the European Parliament Committee on Legal Affairs.

Eden, A.H., E. Steinhart, D. Pearce and J.H. Moor (2012), 'Singularity hypotheses: An overview', in A.H. Eden, J.H. Moor, J.H. Soraker and E. Steinhart (eds), *Singularity Hypotheses: A Scientific and Philosophical Assessment*, Berlin: Springer, pp. 1–12.

Elfes, A. (1990), 'Occupancy grids: A stochastic spatial representation for active robot perception', Paper presented to the Sixth Conference on Uncertainty in AI in Cambridge, MA.

Ellis, E.S. (1868 [2016]), *The Steam Man of the Prairies*, New York, NY: Dover Publications.

Etzkowitz, H. and L. Leydesdorff (1995), 'The Triple Helix: University–industry–government relations – A laboratory for knowledge-based economic development', *EASST Review*, **14** (1), 14–19.

Fagnant, D.J. and K. Kockelman (2015), 'Preparing a nation for autonomous vehicles: Opportunities, barriers and policy recommendations', *Transportation Research Part A*, **77**, 167–181.

Felker, G. (2009), 'The political economy of Southeast Asia's technoglocalism', *Cambridge Review of International Affairs*, **22**, 469–491.

Feng, A. and G. Graetz (2015), 'A question of degree: The effects of degree class on labor market outcomes', IZA Discussion Papers No 8826.

Freeman, C. and F. Louca (2001), *As Time Goes By: From the*

Industrial Revolution to the Information Revolution, Oxford: Oxford University Press.

Frey, C.B. and M.A. Osborne (2017), 'The future of employment: How susceptible are jobs to computerisation?', *Technological Forecasting and Social Change*, **114**, 254–280.

Gadd, M. and P. Newman (2015), 'A framework for infrastructure-free warehouse navigation', Paper presented at the 2015 IEEE International Conference on Robotics and Automation (ICRA) in Seattle, WA, May.

Goldin, C. and L.F. Katz (1998), 'The origins of technology–skill complementarity', *Quarterly Journal of Economics*, **113** (3), 693–732.

Goldin, C. and K. Sokoloff (1982), 'Women, children, and industrialization in the early republic: Evidence from the manufacturing censuses', *The Journal of Economic History*, **42** (4), 741–774.

Goos, M., A. Manning and A. Salomons (2010), 'Explaining job polarization in Europe: The roles of technology, globalization and institutions', Centre for Economic Performance Discussion Papers No 1026.

Graetz, G. and G. Michaels (2017), 'Robots at work', CEP Discussion Paper No 1335.

Granosik, G. (2014), 'Hypermobile robots: The survey', *Journal of Intelligent & Robotic Systems*, **75** (1), 147–169.

Grunwald, A. (2016), 'Societal risk constellations for autonomous driving: Analysis, historical context and assessment', in M. Maurer, J.C. Gerdes, B. Lenz and H. Winner (eds), *Autonomous Driving*, Berlin: Springer, pp. 641–663.

Guilford, T., S. Roberts, D. Biro and I. Rezek (2004), 'Positional entropy during pigeon homing II: Navigational interpretation of Bayesian latent state models', *Journal of Theoretical Biology*, **227** (1), 25–38.

Guiochet, J., M. Machin and H. Waeselynck (2017), 'Safety-critical advanced robots: A survey', *Robotics and Autonomous Systems*, **94**, 43–52.

Hacklin, F., C. Marxt and F. Fahrni (2009), 'Coevolutionary cycles of convergence: An extrapolation from the ICT industry', *Technological Forecasting and Social Change*, **76** (6), 723–736.

Hémous, D. and M. Olsen (2014), 'The rise of the machines: Automation, horizontal innovation and income inequality', IESE Business School Working Paper No WP1110-E.

Henderson, J.V. (2007), 'Understanding knowledge spillovers', *Regional Science and Urban Economics*, **37**, 497–508.

Hitomi, K. (1994), 'Automation: Its concept and a short history', *Technovation*, **14** (2), 121–128.

Hoeckelmann, M., I.J. Rudas, P. Fiorini, F. Kirchner and T. Haidegger (2015), 'Current capabilities and development potential in surgical robotics', *International Journal of Advanced Robotic Systems*, **12** (5), 61–99.

Hornyak, T.N. (2006), *Loving the Machine: The Art and Science of Japanese Robots*, Tokyo: Kodansha International.

Howard, D. and D. Dai (2014), 'Public perceptions of self-driving cars: The case of Berkeley, California', Paper presented at the 93rd Annual Meeting Transportation Research Board, Washington, DC.

Hudson, J. (1988), *Unemployment After Keynes: Towards a New General Theory*, Basingstoke, UK: Palgrave Macmillan.

Hudson, J. and H.F. Khazragui (2013), 'Into the valley of death: Research to innovation', *Drug Discovery Today*, **18** (13–14), 610–613.

Hudson, J. and M. Orviska (2011), 'European attitudes to gene therapy and pharmacogenetics', *Drug Discovery Today*, **16** (19–20), 843–847.

Hudson, J., M. Orviska and J. Hunady (2017a), 'People's attitudes to robots in caring for the elderly', *International Journal of Social Robotics*, **9** (2), 199–210.

Hudson, J., M. Orviska and J. Hunady (2017b), 'Woman vs. machine? Empirical evidence on the effect of robots on employment and prosperity', unpublished paper.

Hudson, J., M. Orviska and J. Hunady (2018), 'People's attitudes to autonomous vehicles', unpublished paper.

Jaffe, A., M. Trajtenberg and R. Henderson (1993), 'Geographic localization of knowledge spillovers as evidenced by patent citations', *Quarterly Journal of Economics*, **63** (3), 577–598.

Janoski, T., D. Luke and C. Oliver (2014), *The Causes of Structural Unemployment: Four Factors that Keep People from the Jobs They Deserve*, London: John Wiley.

Joy, B. (2000), 'Why the future doesn't need us', Wired, https://www.wired.com/2000/04/joy-2/.

Kalra, N. and S.M. Paddock (2016), 'Driving to safety: How many miles of driving would it take to demonstrate autonomous vehicle reliability?', *Transportation Research Part A*, **94**, 182–193.

Katrakazas, C., M. Quddus, W.H. Chen and L. Deka (2015), 'Real-time motion planning methods for autonomous on-road driving: State-of-the-art and future research directions', *Transportation Research Part C*, **60**, 416–442.

Katz, L.F. and R.A. Margo (2013), 'Technical change and the relative demand for skilled labor: The United States in historical perspective', NBER Working Paper No 18752.

Katz, L. and K. Murphy (1992), 'Changes in relative wages, 1963–1987: Supply and demand factors', *Quarterly Journal of Economics*, **107**, 35–78.

Kaushik, D., R. High, C.J. Clark and C.A. LaGrange (2010), 'Malfunction of the Da Vinci robotic system during robot-assisted laparoscopic prostatectomy: An international survey', *Journal of Endourology*, **24** (4), 571–575.

Kellenbenz, H. (1974), 'Technology in the age of the scientific revolution, 1500–1700', *The Fontana Economic History of Europe*, **2**, 177–272.

Keynes, J.M. (1933), 'Economic possibilities for our grandchildren (1930)', in J.M. Keynes, *Essays in Persuasion*, New York, NY: W.W. Norton, pp. 358–373.

Koestler, A. (1964), *The Act of Creation*, New York, NY: Penguin Books.

Korinek, A. (2018), 'The race between human and artificial intelligence', Paper presented at the INET/IMF Conference on the Economics of Artificial Intelligence in Washington, DC.

Krusell, P., L. Ohanian, J.-V. Rios-Rull and G. Violante (2000), 'Capital–skill complementarity and inequality: A macroeconomic analysis', *Econometrica*, **68**, 1029–1053.

Kuipers, B. and Y.T. Byun (1991), 'A robot exploration and mapping strategy based on a semantic hierarchy of spatial representations', *Robotics and Autonomous Systems*, **8** (1–2), 47–63.

Kulshrestha, J., M. Eslami, J. Messias, M.B. Zafar, S. Ghosh, K.P. Gummadi and K. Karahalios (2017), 'Quantifying search bias: Investigating sources of bias for political searches in social media', Paper presented at the ACM Conference on Computer Supported Cooperative Work and Social Computing in Portland, OR.

Kurzweil, R. (2005), *The Singularity is Near: When Humans Transcend Biology*, New York, NY: Viking.

Kyriakidis, M., R. Happee and J.C.F. De Winter (2015), 'Public opinion on automated driving: Results of an international questionnaire among 5000 respondents', *Transportation Research Part F*, **32**, 127–140.

Langlois, A. (2017), 'The global governance of human cloning: The case of UNESCO', *Palgrave Communications*, **3**, 1–8.

Lei, D.T. (2000), 'Industry evolution and competence development: The imperatives of technological convergence', *International Journal of Technology Management*, **19** (7–8), 699–738.

Lenoir, T. (1997), *Instituting Science: The Cultural Production of Scientific Disciplines*, Stanford, CA: Stanford University Press.

Leontief, W. (1983), 'National perspective: The definition of problems and opportunities', in *The Long-Term Impact of Technology on Employment and Unemployment: A National Academy of Engineering Symposium*, Washington, DC: National Academy Press, pp. 3–7.

Levy, F. and R.J. Murnane (2004), *The New Division of Labor: How Computers are Creating the Next Job Market*, Princeton, NJ: Princeton University Press.

Lin, P., K. Abney and G. Bekey (2011), 'Robot ethics: Mapping the issues for a mechanized world', *Artificial Intelligence*, **175** (5–6), 942–949.

Lojek, B. (2007), *History of Semiconductor Engineering*, New York, NY: Springer.

Maloney, W.F. and C. Molina (2016), 'Are automation and trade polarizing developing country labor markets, too?', World Bank Policy Research Working Paper No 7922.

Marchant, G.E., B. Allenby, R.C. Arkin, J. Borenstein, L.M. Gaudet, O. Kittrie and J. Silberman (2015), 'International governance of autonomous military robots', in K. Valavanis and G.J. Vachtsevanos (eds), *Handbook of Unmanned Aerial Vehicles*, Dordrecht, the Netherlands: Springer, pp. 2879–2910.

Marothiya, P. and S.K. Saha (2005), 'Robot inverse kinematics and dynamics algorithms for windows', in *Proceedings of the Conference on Advances and Recent Trends in Manufacturing*, West Bengal: Kalyani Govt College, pp. 229–237.

McCarthy, J. (1977), 'Epistemological problems of artificial intelligence', in *Proceedings of the Fifth International Joint Conference on Artificial Intelligence*, Cambridge, MA: MIT Press, pp. 1038–1044.

Mersky, A.C. and C. Samaras (2016), 'Fuel economy testing of autonomous vehicles', *Transportation Research Part C: Emerging Technologies*, **65**, 31–48.

MGI (2013), 'Disruptive technologies: Advances that will transform life, business, and the global economy', McKinsey Global Institute Technical Report.

Mokyr, J. (1990), *The Lever of Riches: Technological Creativity and Economic Progress*, Oxford: Oxford University Press.

Mokyr, J., C. Vickers and N.L. Ziebarth (2015), 'The history of technological anxiety and the future of economic growth: Is this time different?', *Journal of Economic Perspectives*, **29** (3), 31–50.

Mondada, F., M. Bonani, F. Riedo, M. Briod, L. Pereyre, P. Rétornaz and S. Magnenat (2017), 'Bringing robotics to formal education: The thymio open-source hardware robot', *IEEE Robotics & Automation Magazine*, **24** (1), 77–85.

Moon A., P. Danielson and H.M. Van der Loos (2012), 'Survey-based discussions on morally contentious applications of interactive robotics', *International Journal of Social Robotics*, **4** (1), 77–96.

Mubin, O., C.J. Stevens, S. Shahid, A. Al Mahmud and J.J. Dong (2013), 'A review of the applicability of robots in education', *Journal of Technology in Education and Learning*, **1**, 1–7.

Nagatani, K., S. Kiribayashi, Y. Okada, K. Otake, K. Yoshida, S. Tadokoro, T. Nishimura, T. Yoshida, E. Koyanagi, M. Fukushima and S. Kawatsuma (2013), 'Emergency response to the nuclear accident at the Fukushima Daiichi Nuclear Power Plants using mobile rescue robots', *Journal of Field Robotics*, **30** (1), 44–63.

Nauwelaers, C. and R. Wintjes (2003), 'Towards a new paradigm for innovation policy?', in B. Asheim, A. Isaksen, C. Nauwelaers and F. Tödtling (eds), *Regional Innovation Policy for Small–Medium Enterprises*, Cheltenham, UK and Northampton, MA: Edward Elgar Publishing, pp. 193–220.

Nelson, R. and E. Phelps (1966), 'Investment in humans, technological diffusion, and economic growth', *American Economic Review*, **56**, 69–75.

Nevejans, N. (2016), 'European civil law rules in robotics', The European Parliament Policy Department for Citizens' Rights and Constitutional Affairs.

Nomura, T., T. Kanda, T. Suzuki and K. Kato (2009), 'Age differences and images of robots: Social survey in Japan', *Interaction Studies*, **10** (3), 374–391.

Nüchter, A. and J. Hertzberg (2008), 'Towards semantic maps

for mobile robots', *Robotics and Autonomous Systems*, **56** (11), 915–926.

OECD (2010), *The OECD Innovation Strategy: Innovation to Strengthen Growth and Address Global and Social Challenges*, Paris: OECD.

Omohundro, S. (2008), 'The basic AI drives', in P. Wang, B. Goertzel and S. Franklin (eds), *Artificial General Intelligence 2008: Proceedings of the First AGI Conference*, Amsterdam: IOS Press, pp. 483–492.

Orviska, M. and J. Hudson (2003), 'Tax evasion, civic duty and the law abiding citizen', *European Journal of Political Economy*, **19** (1), 83–102.

Orviska, M., J. Hunady, D. Mlynarova and J. Hudson (2018), 'The role of standards in maximizing the impact and minimizing the dangers of robots', Paper presented at the 23rd EURAS Annual Standardisation Conference in Dublin, Ireland.

Orwell, G. (1949), *Nineteen Eighty-Four: A Novel*, London: Secker & Warburg.

Pak, Y.E. and W. Rhee (2016), 'Convergence science and technology at Seoul National University', in W.S. Bainbridge and M.C. Roco (eds), *Handbook of Science and Technology Convergence*, Cham, Switzerland: Springer International Publishing, pp. 985–1006.

Payre, W., J. Cestac and P. Delhomme (2014), 'Intention to use a fully automated car: Attitudes and a priori acceptability', *Transportation Research Part F*, **27**, 252–263.

Peláez, A.L. and D. Kyriakou (2008), 'Robots, genes and bytes: Technology development and social changes towards the year 2020', *Technological Forecasting and Social Change*, **75** (8), 1176–1201.

Phua, C., V. Lee, K. Smith and R. Gayler (2010), 'A comprehensive survey of data mining-based fraud detection research', arXiv pre-print arXiv:1009.6119.

Polanyi, M. (1966), *The Tacit Dimension*, New York, NY: Doubleday.

Porpiglia, F., I. Morra, M.L. Chiarissi, M. Manfredi, F. Mele, S. Grande, F. Ragni, M. Poggio and C. Fiori (2013), 'Randomised controlled trial comparing laparoscopic and robot-assisted radical prostatectomy', *European Urology*, **63** (4), 606–614.

Pratt, G.A. (2015), 'Is a Cambrian explosion coming for robotics?', *Journal of Economic Perspectives*, **29** (3), 51–60.

Reed, H. and S. Lansley (2016), *Universal Basic Income: An Idea Whose Time has Come?* London: Compass.

Reiss, M.J. (1999), 'Teaching ethics in science', *Studies in Science Education*, **34** (1), 115–140.

Rogers, E. (1995), *Diffusion of Innovations*, 4th edn, New York, NY: Free Press.

Rosenthal, S.S. and W.C. Strange (2004), 'Evidence on the nature and sources of agglomeration economies', in J.V. Henderson and J.F. Thisse (eds), *Handbook of Regional and Urban Economics, Vol. 4*, Amsterdam: North Holland, pp. 2119–2171.

Roubini, N. (2014), 'Rise of the machines: Downfall of the economy?', http://www.economonitor.com/nouriel/2014/12/08/rise-of-the-machines-downfall-of-the-economy/ (accessed 14 April 2018).

Sachs, J.D. (2018), 'Structural transformation, and the distribution of income', Paper presented at the INET/IMF Conference on the Economics of Artificial Intelligence in Washington, DC.

Sachs, J.D, S.G. Benzell and G. LaGarda (2015), 'Robots: Curse or blessing? A basic framework', NBER Working Paper No 21091.

Sandberg, A. (2010), 'An overview of models of technological singularity', Paper presented at the Third Conference on Artificial General Intelligence, Lugano, Switzerland.

Sarwar, M. and T.R. Soomro (2013), 'Impact of smartphones on society', *European Journal of Scientific Research*, **98** (2), 216–226.

Schmitt, J., H. Schierholz and L. Mishel (2013), 'Don't blame the robots: Assessing the job polarization explanation of growing wage inequality', Economic Policy Institute Report, https://www.epi.org/publication/technology-inequality-dont-blame-the-robots/.

Schoettle, B. and M. Sivak (2014), 'A survey of public opinion about autonomous and self-driving vehicles in the U.S., the U.K., and Australia', University of Michigan Transportation Research Institute Report No 2014-21.

Searle, J.R. (1980), 'Minds, brains, and programs', *Behavioral and Brain Sciences*, **3** (3), 417–424.

Senarens, L. (1876 [1893]), *Frank Reade and his Electric Man*, Frank Reade Library, Volume 2, Issue 37, New York, NY: Frank Tousey.

Sharkey, A.J. (2016), 'Should we welcome robot teachers?', *Ethics and Information Technology*, **18** (4), 283–297.

Sharkey, N., M. Goodman and N. Ross (2010), 'The coming robot crime wave', *Computer*, **43** (8), 115–116.

Shelley, M. (1818), *Frankenstein; Or, The Modern Prometheus*, London: Lackington, Hughes, Harding, Mavor & Jones.

Speak, S. (2000), 'Barriers to lone parents' employment: Looking beyond the obvious', *Local Economy*, **15** (1), 32–44.

Spence, M. (1984), 'Cost reduction, competition, and industry performance', *Econometrica*, **52**, 101–121.

Stiglitz, J. (2018), 'Lecture on the macroeconomics of AI', Paper presented at the INET/IMF Conference on the Economics of Artificial Intelligence in Washington, DC.

Stone, W.L. (2005), 'The history of robotics', in T.R. Kurfess (ed.), *Robotics and Automation Handbook*, New York, NY: CRC Press, pp. 1–12.

Straub, E.T. (2009), 'Understanding technology adoption: Theory and future directions for informal learning', *Review of Educational Research*, **79** (2), 625–649.

Sturgis, P. and N. Allum (2004), 'Science in society: Re-evaluating the deficit model of public attitudes', *Public Understanding of Science*, **13** (1), 55–74.

Taipale S., F. de Luca, M. Sarricaand and L. Fortunati (2015), 'Robot shift from industrial production to social reproduction', in J. Vincent, S. Taipale, B. Sapio, G. Lugano and L. Fortunati (eds), *Social Robots from a Human Perspective*, Berlin: Springer, pp. 11–24.

Tan, A., H. Ashrafian, A.J. Scott, S.E. Mason, L. Harling, T. Athanasiou and A. Darzi (2016), 'Robotic surgery: Disruptive innovation or unfulfilled promise? A systematic review and meta-analysis of the first 30 years', *Surgical Endoscopy*, **30** (10), 4330–4352.

Tanaka, F. and T. Kimura (2010), 'Care-receiving robot as a tool of teachers in child education', *Interaction Studies*, **11** (2), 263–268.

Theodoridis, T. and H. Hu (2012), 'Toward intelligent security robots: A survey', *IEEE Transactions on Systems, Man, and Cybernetics, Part C (Applications and Reviews)*, **42** (6), 1219–1230.

Todtling, F. and M. Trippl (2005), 'One size fits all? Towards a differentiated regional innovation policy approach', *Research Policy*, **34**, 1203–1219.

Turing, A.M. (1950), 'Computing machinery and intelligence', *Mind*, **59** (236), 433–460.

Turing, A.M. (1951), 'Intelligent machinery: A heretical theory', The 51 Society, BBC programme.

Turkle, S. (2011), *Alone Together: Why We Expect More from Technology and Less from Each Other*, New York, NY: Basic Books.

Villiers de l'Isle-Adam, V. (1886 [1993]), *L'Ève future*, Paris: Folio.

Wadud, Z., D. MacKenzie and P. Leiby (2016), 'Help or hindrance? The travel, energy and carbon impacts of highly automated vehicles', *Transportation Research Part A*, **86**, 1–18.

Wolmar, C. (2009), *Fire and Steam: A New History of the Railways in Britain*, London: Atlantic Books.

World Bank (2008), *Global Economic Prospects 2008: Technology Diffusion in the Developing World*, Washington, DC: World Bank.

World Bank (2010), *World Development Report 2010: Development and Change*, Washington, DC: World Bank.

World Bank (2016), *World Development Report 2016: Digital Dividends*, Washington, DC: World Bank.

Index

160   *The robot revolution*

Sweden 23, 28, 73, 108–9, 112, 114
Switzerland 28

Taiwan 28
technology adoption 9
Tiro 39
Tokyo Institute of Technology 45
Turing, Alan 61, 94
Turing test 61

Uganda, sweet potato 4, 5
UK 2, 7, 12, 21, 23, 28, 63, 102, 109,
    115, 119, 136–7
unemployment 10, 12, 17, 47,
    63–86, 88–90, 111–13, 117,
    121–2, 131–2, 137
United Nations (UN) 43, 126
universities, role of 6, 15, 22–4,
    39–40, 42–3, 45–6, 118–19,
    121, 128, 134–9, 142–3
University College London
    (UCL)
unmanned aerial vehicles (UAV)
    42–3, 50

unmanned ground vehicles (UGV)
    42, 44
USA 2, 14, 27–8, 34, 38, 39, 42–3,
    45, 63, 65, 68, 99, 126, 129,
    137

vacuum cleaner 2
vecro, discovery of 6
vehicles, driverless/autonomous
    35–7, 53, 67, 96–7, 101–2,
    105–6, 108–10, 112, 113–14,
    116, 131
Villiers, Auguste 25

WABOT-1 23
WABOT-2 23
Waseda University 23
Watt, James 136
World Bank 8–9
Wozniak, Steve 126

Yale University 137

Zuckerberg, Mark 122